Springer Theses

Recognizing Outstanding Ph.D. Research

Aims and Scope

The series "Springer Theses" brings together a selection of the very best Ph.D. theses from around the world and across the physical sciences. Nominated and endorsed by two recognized specialists, each published volume has been selected for its scientific excellence and the high impact of its contents for the pertinent field of research. For greater accessibility to non-specialists, the published versions include an extended introduction, as well as a foreword by the student's supervisor explaining the special relevance of the work for the field. As a whole, the series will provide a valuable resource both for newcomers to the research fields described, and for other scientists seeking detailed background information on special questions. Finally, it provides an accredited documentation of the valuable contributions made by today's younger generation of scientists.

Theses may be nominated for publication in this series by heads of department at internationally leading universities or institutes and should fulfill all of the following criteria

- They must be written in good English.
- The topic should fall within the confines of Chemistry, Physics, Earth Sciences, Engineering and related interdisciplinary fields such as Materials, Nanoscience, Chemical Engineering, Complex Systems and Biophysics.
- The work reported in the thesis must represent a significant scientific advance.
- If the thesis includes previously published material, permission to reproduce this must be gained from the respective copyright holder (a maximum 30% of the thesis should be a verbatim reproduction from the author's previous publications).
- They must have been examined and passed during the 12 months prior to nomination.
- Each thesis should include a foreword by the supervisor outlining the significance of its content.
- The theses should have a clearly defined structure including an introduction accessible to new PhD students and scientists not expert in the relevant field.

Indexed by zbMATH.

More information about this series at https://link.springer.com/bookseries/8790

Jannis Lehmann

Toroidal Order in Magnetic Metamaterials

Doctoral Thesis accepted by
ETH Zurich, Zurich, Switzerland

Springer

Author
Dr. Jannis Lehmann
Laboratory for Multifunctional
Ferroic Materials
Department of Materials
ETH Zurich
Zurich, Switzerland

Supervisor
Prof. Dr. Manfred Fiebig
Laboratory for Multifunctional Ferroic
Materials
Department of Materials
ETH Zurich
Zurich, Switzerland

ISSN 2190-5053 ISSN 2190-5061 (electronic)
Springer Theses
ISBN 978-3-030-85494-2 ISBN 978-3-030-85495-9 (eBook)
https://doi.org/10.1007/978-3-030-85495-9

This Springer imprint is published by the registered company Springer Nature Switzerland AG
The registered company address is: Gewerbestrasse 11, 6330 Cham, Switzerland

Supervisor's Foreword

The property of materials exhibiting spontaneous forms of long range, so-called ferroic, order is one of the most fundamental phenomena exhibited by condensed matter systems. Ferromagnetism as spontaneous order of magnetic moments has been known to humankind for millennia and is of immense technological value. Ferroelectricity as spontaneous order of electric dipole moments was discovered exactly a century ago, and ferroelasticity as spontaneous deformation of a material was recognized as a ferroic form of ordering around 1970.

Currently, a variety of candidates for "novel" types of ferroic order are being discussed. This is foremost ferrotoroidicity, the spontaneous uniform order of magnetic whirls which may complement antiferromagnetism as alternative "magnetization-free" magnetic state. The lack of a magnetization can be of particular advantage when exploring fast magnetic switching processes and magnetoelectronic applications. Despite the interest in such novel types of ferroics, their study is difficult because a magnetic order without magnetic field is not easy to address experimentally and associated concepts like "toroidal magnetic fields" remain largely elusive.

In his doctoral thesis, Jannis Lehmann describes how he explores magnetization-free manifestations of magnetic order, improves their conceptual understanding and demonstrates read/write access to these in externally applied fields. To achieve this, he develops a fundamentally new approach. Instead of working with atomic matter, where crystals are composed of unit cells with lateral extensions on the order of 1 nm, he resorts to the use of artificial crystals in the form of two-dimensional arrays assembled from rod-like nanomagnets. The "atoms" in these arrays are represented by thin permalloy stripes with a length on the order of 0.1–1 μm. These stripes display long-range magnetic order stabilized by the magnetic dipole rather than by the exchange interaction. Because of the lateral upscaling, the local magnetic order and properties of the artificial crystals are, in contrast to their natural equivalents, accessible to scanning probe microscopy.

At first, Jannis explores the intrinsic mechanisms determining the magnetic arrangement of the nanoarrays. His experiments show that they stabilize a spontaneously long-range ordered state, including the formation of domains to which an order parameter can be associated. He designs these systems in micromagnetic simulations and in close contact with the lithography team such that only two interactions as classical equivalents of the symmetric and antisymmetric exchange interaction determine the resulting magnetic order. By engineering his toy system to exhibit zero net magnetization, intrinsic mechanisms determining the size and shape of domains come to the fore. In particular, Jannis shows how size and shape are determined by the competition of the aforementioned "classical exchange contributions".

In a second, more specific approach, Jannis examines the "hottest" new member of the family of established types of ferroic order, the aforementioned ferrotoroidicity. He presents arrays with an inherent preference for a ferrotoroidic ground state. Most importantly, he uncovers the nature of the toroidal magnetic field and demonstrates toroidal field poling. In still ongoing experiments, Jannis searches for unusual optical effects displayed by the ferrotoroidic arrays. Their specific symmetry is expected to lead to a difference between forward- and backward-propagating light, which could even be the basis of an optical diode.

In summary, the doctoral thesis of Jannis Lehmann has initiated a substantial advance in our understanding and concept of magneto-toroidal order with consequences for ferroic systems at large. In particular, the idea to upscale the magnetocrystalline lattice to explore hidden types of magnetic long-range order should be of lasting value.

Zurich, Switzerland
March 2021

Prof. Dr. Manfred Fiebig

Abstract

While often materials arrange themselves in a plain and symmetric way, it is the absence of symmetry that causes a variety of fascinating phenomena. Ferroic materials, for example, exhibit a long-range order that results from a spontaneous symmetry breaking. This order is associated with a macroscopic tensorial property that can be oriented by the application of a so-called conjugate field. These key characteristics are the foundation of the relevance and broad use of ferroics in current scientific research and in technological applications. While some prototypic ferroic materials such as ferromagnetic iron have been used by mankind for millennia and are microscopically well understood, recently proposed new types of ferroic order are still in their infancy. Ferrotoroidicity, characterized by a simultaneous violation of both space-inversion and time-reversal symmetry, is an interesting candidate among these. Microscopically, this type of order is based on a compensated alignment of magnetic moments that form whirls at the level of a unit cell, so-called toroidal moments. An inherent linear coupling of magnetic and electric properties is accompanied by a collective alignment of toroidal moments, which promises exotic electric, magnetic and optical responses. However, detection and control of ferrotoroidic order are experimentally challenging. On the one hand, the zero net magnetization of the ferrotoroidic state hinders a sensing by conventional magnetic probing techniques. On the other hand, the generation of a conjugate field to deliberately manipulate the order, for instance a magnetic vortex field at the level of a unit cell, appears elusive. As a consequence, profound experimental knowledge about this subtle ferroic state is still lacking. These circumstances motivate new approaches for scrutinizing ferrotoroidicity.

The scope of this work is to study ferrotoroidicity by transferring basic spin configurations pertinent for the emergence of toroidal order to mesoscopic length scales. The use of two-dimensional nanomagnetic arrays allows for engineering and accessing the system's relevant magnetic degrees of freedom. This approach enables to elucidate microscopic and macroscopic aspects of toroidally ordered matter beyond the reach of natural materials. I demonstrate the design of planar arrays comprised of sub-micrometre-sized nanomagnets that are coupled by well-defined and tunable magnetostatic interactions and display a macroscopically ordered ferrotoroidic state. For addressing fundamental considerations about ferrotoroidic order in the upscaled system, I identify three milestones:

1. The tunability of spontaneously formed toroidal domains,
2. the local control over the toroidal order parameter, and
3. the detection of optical signatures related to the toroidal order.

As part of my work, I extricated intrinsic coupling parameters that mediate the transition to long-range order in artificial ferroic crystals. By modifying the interplay of short-range interactions, the size and morphology of as-grown ferrotoroidic domains could be controlled. A simultaneous transition between different realizations of domain-wall structures was found. Further, I developed a new method for locally manipulating the order parameter by scanning an artificial ferrotoroidic array with the magnetic tip of a scanning probe microscope. This strategy allowed for "sequentially activating" crystal magnetic fields that originate from superimposed stray fields of surrounding nanomagnets. Unit-cell-scale magnetic vortex fields of either sign have been applied to achieve local control over the orientation of toroidal moments. For detecting optical signatures that originate from the toroidal order, considerations of the magnetic arrays representing two-dimensional gratings allowed to apply a novel magneto-optical diffraction technique. This technique revealed a directional anisotropy that may be attributed to the toroidal state.

The investigations for my doctorate extend the knowledge about ferrotoroidicity and provide a basis for future research on materials in which a certain type of magnetic order plays an essential role. Further, the concepts of transferring microscopic interactions, ordering schemes and manipulation methods to tangible length scales without loosing the basic macroscopic behaviour may stimulate analogous approaches in other fields of research touched by novel ordering phenomena or complex types of magnetism.

A consecutive aspect of the upscaling from atomic dimensions towards more accessible length scales has been touched in the context of educational methods in applied crystallography. Here, for providing a more intuitive access to crystal diffraction, I developed an optical simulation based on a modified video projector. It allows for interactively demonstrating crystal diffraction in the optical regime in a classroom. Furthermore, I suggested and outlined a new laboratory course at the Department of Materials at ETH Zurich, in which I built and used a data acquisition system based on a computer sound card to measure, for instance, frequency-dependent electrical impedance curves.

Acknowledgements

I want to say *thank you* to a couple of people who helped and accompanied me by getting where I am and who I am today. I begin with the examiner board of my doctorate: first, I want to thank my advisor Manfred Fiebig for giving me the opportunity to work on my doctorate in his nicely developing and growing group at ETH Zurich. Working under his supervision gave me insight into fields of physics and materials science that were new to me and extended my knowledge. Furthermore, he gave me an idea of the nowadays publishing system and, during extended periods of manuscript writings, he taught me the important skill of compressing a topic with all its technical and nerdy details (that usually drives and motivates scientists like me) to its comprehensible essence. He allowed me to work very independent and trusted in my instinct to choose meaningful scientific directions for my project. In addition, I am happy that I had his support for participating in scientific meetings in Europe, Asia and America. I further appreciated having the duty to teach several exercise and laboratory courses with bachelor students as well as the freedom to do some side research such as the "from scratch"–construction of set-ups for teaching crystallography (that eventually resulted in a publication, live demos at several courses at ETH and at an international conference) and for "do-it-yourself" data acquisition (envisaged for a laboratory course at ETH Zurich). Second, my gratitude goes to Laura Heyderman for developing the idea of nanomagnetic toroidal systems in collaboration with Manfred Fiebig, for supporting me and my research and for helping me on scientific and formal questions during my time as a doctorate student. Discussions with her and her group members were always helpful for getting a different perspective and background for my results. Third, I want to thank both Maxim Mostovoy and Rolf Allenspach for evaluating my work and for serving as very suitable external referees with a strong background on research fields close to my work. Another "thank you" goes to Jörg Löffler for being the chair for my dissertation and disputation.

Furthermore, I want to thank all the former and current members of the ferroic group for providing a great atmosphere to work in. In particular, I would like to mention:

Our secretary Ursula Bruzzone for all her help with administrative issues as well as Josef Hecht and Sebastian Reitz for their support in our tool shop. Peggy Schoenherr and Martin Lilienblum for introducing me to scanning probe techniques. Thomas

Lottermoser for introducing Oktay Aktas, Christian Tzschaschel and me to nonlinear optical imaging and spectroscopy as well as for all the valuable discussions about manuscripts. Oktay Aktas and Christian Tzschaschel for providing a great team, in particular in the beginning of our shared time at ETH Zurich. Stefan Günther for extended discussions and brainstorming sessions about ultrafast spectroscopy and, further, for introducing me to Labview and for helping in the design and construction of the programmable magnet. Mads Weber, Shovon Pal and Morgan Trassin for pushing me when necessary and for always welcoming discussions about how do to that, how to write that, how to present that, how to complain about that and lots of unscientific stuff. Lukas Kürten and Julia Knapp for providing the nice template base for my LaTeX document. Claire Donnelly for growing my samples throughout the first two years of my doctorate as well as for her help in writing our first manuscript, preparing figures and for drinking mulled wine at Christmas times. Naëmi Leo for growing a couple of nice samples during the third year of my doctorate as well as for scientific and non-scientific discussions and great alpine hikes. Leonardo Degiorgi and his group at ETH Zurich for performing FTIR spectroscopy on some of my samples. Amadé Bortis, Lukas Kürten, Thomas Lottermoser, Mads Weber and Manfred Fiebig for proofreading parts of this thesis and Dumeni Manatschal for his work on a recognition code to determine the magnet configuration in my samples.

At the end, I want to express my thanks to the persons close to me who shared happy as well as challenging times. These are in particular my dear friends from Hamburg, Freiburg and Zurich, Katja Lell, Lotti, Meune, Hendrik and Niko Lehmann and Wiebi, Levin and Olivia Herzberg. You are great ♡

Contents

Chapter 1
Introduction

In the formation of ice flowers after breathing against a cold window pane, nature impressively displays its tendency to spontaneously order in well-defined symmetric configurations. The example depicts an elementary feature of phase transitions [1] from the disordered phase of water vapour, to the short-range-ordered phase present in the condensed liquid water into the final long-range-ordered phase of frozen ice crystals. Here, during the transition from a disordered to an ordered phase, the breaking of symmetries defines physical properties and macroscopic characteristics of the final phase.

The concept of order that emerges from a spontaneous symmetry breaking is a fundamental aspect of physics and extends into a large variety of phenomena beyond the well-known changes between aggregate states as introduced above. An order in the context of solid-state physics is often related to the collective alignment of local constituents of a considered system. Cooperative effects between charges and magnetic moments in a solid are, for example, key for the description of strongly correlated electronic systems [2] such as superconductors [3], charge- and spin-density waves [4] and in particular for understanding ferroic materials [5, 6]. Common characteristic of these ferroic materials is their preference to spontaneously develop areas across which spins and/or charges align uniformly in well-defined configurations, known as ferroic domains. The spatial orientation of the local order in these areas can be manipulated at will by the application of a particular, so-called conjugate field. An example for a ferroic state that is of tremendous importance in the cogwheel of modern technology is the parallel alignment of magnetic moments that macroscopically break time-reversal symmetry in ferromagnetic materials such as iron. The magnetisation and the magnetic field associated with ferromagnetism are inevitable in our daily life as they provide the basis of power generation and transformation, electric motors, sensors, digital memory grids, and much more [7–9]. In contrast to ferromagnetic materials, there are other types of ferroic order that are much less tangible and consequently not so familiar to the general public, such as

J. Lehmann, *Toroidal Order in Magnetic Metamaterials*, Springer Theses,
https://doi.org/10.1007/978-3-030-85495-9_1

ferroelectric materials, with a spontaneous alignment of electric dipole moments, or ferroelastic materials with a spontaneous mechanical strain.

Moreover, there are different types of *magnets* that, for instance, despite their internal order, neither possess a net magnetisation nor reorient in an external magnetic field. The magnetic moments in these so-called compensated magnets are arranged in such way that the sum over the individual fields vanishes. Despite, but at the same time owing to, the non-existent magnetization and magnetic field, there are technologically beneficial properties associated to those compounds. Compensated types of magnetic order are more robust against external perturbations, exhibit a very fast response of the spin system [8, 10] and are the basis for phenomena such as magnetically induced ferroelectricity [11–13], exchange bias [14, 15], or certain magnetoresistance effects [16]. Hence, these types of magnetic order are considered to provide the basis for an advanced memory and spintronic technology [10, 17, 18]. If a certain arrangement of magnetic moments leads to a macroscopic space- and time-reversal symmetry breaking, an intrinsic linear magnetoelectric effect is allowed to arise [19]. As a consequence, such a state of matter would be particularly useful for manipulating magnetic properties by electric fields and vice versa. Different arrangements of magnetic moments have been proposed that facilitate such symmetry requirements for a novel type of ferroic state. Besides the suggested monopolar and quadrupolar magnetic order [20, 21], a toroidal order [22–25] stands out as a very suitable candidate fulfilling hallmark properties of ferroic states—spontaneous domain formation and conjugate-field poling [26–28]. Ferrotoroidic order consists of unit-cell-sized whirls of magnetic moments that collectively wind either clockwise or counter-clockwise. This fascinating type of compensated magnetic order has attracted great interest because of its intrinsic linear magnetoelectric nature [24]. This allows for magnetoelectric sensors or data-storage applications in which ferrotoroidic bits may replace ferromagnetic bits in an advanced robust and fast computer memory [29]. In that aspect, ferrotoroidics display their proximity to magnetoelectric multiferroic materials in which magnetic and electric orders coexist and are possibly interconnected [21, 30–33].

Ferrotoroidicity is the least developed member in the group of known ferroic states, otherwise including ferroelastics, ferroelectrics and ferromagnets. This is primarily due to experimental challenges that result from the elusiveness of accessing and controlling the corresponding order parameter. Hence, ferrotoroidic candidate materials can be identified merely indirectly via a symmetry analysis of the magnetic point group and by probing certain components of the magnetoelectric interaction tensor. In order to circumvent these experimental obstacles, a new fruitful approach is presented in this work.

Artificial planar nanostructures made from ferromagnetic building blocks provide a versatile platform for the study of fundamental phenomena in magnetism such as frustration, ordering and relaxation [34–37], in which responses can be tuned via size, shape, material, arrangement and symmetry of the constituents. Advanced techniques to nanoscopically pattern materials on demand by lithographic methods allow for tailoring those systems to address further scientific questions, e.g., by imitating a particular macroscopic behaviour of spin systems. For the purpose of this work, arrays of stray-field-coupled nanomagnets in the ferromagnetic single-domain- or vortex-state have been designed and fabricated in different geometries. By patterning magnetic building blocks in arrangements that match with a ferrotoroidic-state symmetry, the phenomena of interest can be transferred and studied at mesoscopic length scales. Here the advantages over a conventional crystalline ferrotoroidic system are twofold. On the one hand, ferroic order exists as a macroscopic phenomenon, independent of its precise microscopic origin. This provides the possibility to mimic quantum-mechanical exchange interactions with suitably implemented classical magnetostatic couplings that are much easier to design and control. On the other hand, the transfer to the mesoscale allows to spatially resolve the structural configuration and provides direct access to magnetic degrees of freedom, using established measurement techniques. Both of these conditions are beneficial for the study of an otherwise inaccessible and elusive type of long-range magnetic order.

In addition, such artificial nanomagnetic arrays with sub-micrometre periodicities are known to act as plasmonic metasurfaces [38–43]. This class of materials has become a vibrant subject of research in the last decades and can provide resonant-field-enhanced electromagnetic responses and associated optical properties beyond the reach of conventional materials.

The goal of my work is to investigate and establish methods to implement and observe ferrotoroidicity in two-dimensional arrays of ferromagnetic building blocks. The main focus lies on the elaboration of key properties of the ferroic state, namely the spontaneous formation of toroidal domains and their poling with a conjugate toroidal field. Here, in contrast to materials with a net magnetisation, domain formation in ferrotoroidic crystals is not necessarily driven by antagonistic contributions to the internal energy, but is mainly due to thermal energy. The key observation of spontaneous domain formation allows for consecutive studies on the microscopic mechanisms that enable to tune the domain configurations—a highly challenging task when approached on the atomic scale. Furthermore, different schemes for the application of a conjugate field to imbalance the different orientations of toroidal moments are conceived and tested. Here, the scaling up to mesoscopic length scales paves a way for the realisation of unconventional field configurations. In addition, the system under consideration motivates the study of optical manifestations of toroidal order as for instance the existence of an optical-diode-like behaviour, namely a non-reciprocal directional anisotropy, directly linked to the orientation of ferrotoroidic domain states. Suitable detection schemes are conceived that utilize optical diffraction resulting from the micrometre-scale periodic patterning of the structure.

The outline of this book is the following:

In Chap. 2 the theoretical background for the main topics is given—ferroic order, nanomagnetism and optical metamaterials.

In Chap. 3 the experimental and computational techniques are introduced that have been applied during the studies. A main focus lies on magnetic force microscopy (MFM) and variants of the magneto-optical Kerr effect (MOKE).

In Chap. 4 the arrangements of magnetic building blocks are classified that allow for the emergence of ferrotoroidic order. Furthermore, experimental challenges and measurement artefacts are explained and options for their circumvention are presented.

In Chap. 5 the formation of short- and long-range order in the systems under consideration is discussed. The Chapter opens with an investigation of domain pinning by applying demagnetisation protocols. The main part deals with the dependence of domain formation on the competition of microscopic couplings and reveals the possibility to tune the ferroic domain size and the domain-wall morphology.

In Chap. 6 different experimental schemes to lift the degeneracy of the two orientations of the toroidisation are discussed. Therefore, three experiments are presented that, first, exploit the local 'activation' of crystal magnetic fields, second, the application of displacement currents and, third, a magnetoelectric interaction with the toroidal state.

In Chap. 7 an experimental method that is based on optical diffraction is introduced that provides sensitivity to particular symmetry violations in nanoscopic periodic arrays. The application of the technique allows for detecting signatures that supposedly correspond to the orientation of the toroidisation and thus provides an access to a compensated magnetic order by linear-optical means.

In Chap. 8 my work on magneto-toroidal metamaterials is concluded and contextualised.

Beyond that, in Appendix A, newly development teaching concepts are presented, in Appendix B the design and construction of laboratory hardware is explained, and the book ends with Appendix C in which additional experimental results are presented.

References

1. Landau LD (1937) On the theory of phase transitions. J Exp Theoret Phys 7:19–32
2. Dagotto E (2005) Complexity in strongly correlated electronic systems. Science 309(5732):257–262. https://doi.org/10.1126/science.1107559
3. Kleiner R, Buckel W (eds) (2016) Superconductivity: an introduction. Wiley-VCH. https://doi.org/10.1002/9783527686513
4. Brown S, Gruener G (1994) Charge and spin density waves. Sci Am 270(4):50–56. https://doi.org/10.1038/scientificamerican0494-
5. Wadhawan VK (2000) Introduction to ferroic materials. 1st edn. CRC Press. https://doi.org/10.1201/9781482283051
6. Tagantsev AK, Cross LE, Fousek J (2010) Domains in ferroic crystals and thin films. Springer, New York. https://doi.org/10.10072F978-1-4419-1417-0
7. Stamps RL et al (2014) The 2014 magnetism roadmap. J Phys D: Appl Phys 47(33):333001. https://doi.org/10.1088/0022-3727/47/33/333001
8. Sander D et al (2017) The 2017 magnetism roadmap. J Phys D: Appl Phys 50(36):363001. https://doi.org/10.1088/1361-6463/aa81a1
9. Vedmedenko EY et al (2020) The 2020 magnetism roadmap. J Phys D: Appl Phys 53(45):453001. https://doi.org/10.1088/1361-6463/ab9d98
10. Gomonay EV, Loktev VM (2014) Spintronics of antiferromagnetic systems (Review Article). Low Temp Phys 40(1):17–35. http://aip.scitation.org/doi/10.1063/1.4862467(104.17.164.62)
11. Mostovoy M (2006) Ferroelectricity in spiral magnets. Phys Rev Lett 96(6):067601. https://doi.org/10.1103/PhysRevLett.96.067601

12. Cheong S-W, Mostovoy M (2007) Multiferroics: a magnetic twist for ferroelectricity. Nat Mater 6(1):13–20. https://doi.org/10.1038/nmat1804
13. Tokura Y, Seki S, Nagaosa N (2014) Multiferroics of spin origin. Rep Prog Phys 77(7):076501. https://doi.org/10.1088/0034-4885/77/7/076501
14. Meiklejohn WH, Bean CP (1957) New magnetic anisotropy. Phys Rev 105(3):904–913. https://doi.org/10.1103/PhysRev.105.904
15. Nogues J, Schuller IK (1999) Exchange bias. J Magnet Magnet Mater 192(2):203–232. https://doi.org/10.1016/S0304-8853(98)00266-2
16. Marti X et al (2014) Room-temperature antiferromagnetic memory resistor. Nat Mater 13(4):367–374. https://www.nature.com/articles/nmat3861
17. Baltz V et al (2018) Antiferromagnetic spintronics. Rev Mod Phys 90(1):015005. https://doi.org/10.1103/RevModPhys.90.015005
18. Jungwirth T et al (2018) The multiple directions of antiferromagnetic spintronics. Nat Phys 14(3):200–203. https://www.nature.com/articles/s41567-018-0063-6
19. Fiebig M (2005) Revival of the magnetoelectric effect. J Phys D: Appl Phys 38(8):123–152 https://iopscience.iop.org/article/10.1088/0022-3727/38/8/R01
20. Spaldin NA et al (2013) Monopole-based formalism for the diagonal magnetoelectric response. Phys Rev B 88(9):094429. https://doi.org/10.1103/PhysRevB.88.094429
21. Spaldin NA, Ramesh R (2019) Advances in magnetoelectric multiferroics. Nat Mater 18(3):203–212. https://www.nature.com/articles/s41563-018-0275-2
22. Dubovik V, Tugushev V (1990) Toroid moments in electrodynamics and solid-state physics. Phys Rep 187(4):145–202. https://www.nature.com/articles/s41563-018-0275-2
23. Ederer C, Spaldin NA (2007) Towards a microscopic theory of toroidal moments in bulk periodic crystals. Phys Rev B 76(21):214404. https://doi.org/10.1103/PhysRevB.76.214404
24. Spaldin NA, Fiebig M, Mostovoy M (2008) The toroidal moment in condensed-matter physics and its relation to the magnetoelectric effect. J Phys: Condensed Matter 20(43):434203. https://doi.org/10.1088/0953-8984/20/43/434203
25. Gnewuch S, Rodriguez EE (2019) The fourth ferroic order: current status on ferrotoroidic materials. J Solid State Chem 271:175–190. https://doi.org/10.1016/j.jssc.2018.12.035
26. Van Aken BB et al (2007) Observation of ferrotoroidic domains. Nature 449(7163):702–705. http://dx.doi.org/10.1038/nature06139
27. Zimmermann AS, Meier D, Fiebig M (2014) Ferroic nature of magnetic toroidal order. Nat Commun 5(1):4796. https://www.nature.com/articles/ncomms5796
28. Toledano P et al (2015) Primary ferrotoroidicity in antiferromagnets. Phys Rev B 92(9):094431. https://doi.org/10.1103/PhysRevB.92.094431
29. Kleemann W (2017) Multiferroic and magnetoelectric nanocomposites for data processing. J Phys D: Appl Phys 50(22):223001. https://iopscience.iop.org/article/10.1088/1361-6463/aa6c04
30. Schmid H (2008) Some symmetry aspects of ferroics and single phase multiferroics. J Phys: Cond Matter 20(43):434201. https://doi.org/10.1088/0953-8984/20/43/434201
31. Pyatakov AP, Zvezdin AK (2012) Magnetoelectric and multiferroic media. Phys Uspekhi 55(6):557–581. https://iopscience.iop.org/article/10.3367/UFNe.0182.201206b.0593
32. Dong S et al (2015) Multiferroic materials and magnetoelectric physics: symmetry, entanglement, excitation, and topology. Adv Phys 64(5):519–626. https://www.tandfonline.com/doi/full/10.1080/00018732.2015.1114338
33. Fiebig M et al (2016) The evolution of multiferroics. Nat Rev Mater 1(8):16046. https://www.nature.com/articles/natrevmats201646
34. Heyderman LJ, Stamps RL (2013) Artificial ferroic systems: novel functionality from structure, interactions and dynamics. J Phys: Condensed Matter 25(36):363201 https://iopscience.iop.org/article/10.1088/0953-8984/25/36/363201
35. Nisoli C, Moessner R, Schiffer P (2013) Colloquium: artificial spin ice: designing and imaging magnetic frustration. Rev Modern Phys 85(4):1473–1490. https://doi.org/10.1103/RevModPhys.85.1473

36. Rougemaille N, Canals B (2019) Cooperative magnetic phenomena in artificial spin systems: spin liquids, Coulomb phase and fragmentation of magnetism—a colloquium. Eur Phys J B 92(3):62. https://doi.org/10.1140/epjb/e2018-90346-7
37. Skjaervo SH et al (2020) Advances in artificial spin ice. Nat Rev Phys 2(1):13–28. https://www.nature.com/articles/s42254-019-0118-3
38. Stockman MI (2011) Nanoplasmonics: the Physics behind the applications. Phys Today 64(2):39–44
39. Hayashi S, Okamoto T (2012) Plasmonics: visit the past to know the future. J Phys D: App Phys 45(43):433001. https://doi.org/10.1088/0022-3727/45/43/433001
40. Alu A (2016) Prime time. Nat Mater 15(12):1229–1231. https://www.nature.com/articles/nmat4814
41. Monticone F, Alu A (2017) Metamaterial, plasmonic and nanophotonic devices. Rep Prog Phys 80(3):036401. https://doi.org/10.1088/1361-6633/aa518f
42. Kravets VG et al (2018) Plasmonic surface lattice resonances: a review of properties and applications. Chem Rev 118(12):5912–5951. https://doi.org/10.1021/acs.chemrev.8b00243
43. Shaltout AM, Shalaev VM, Brongersma ML (2019) Spatiotemporal light control with active metasurfaces. Science 364(6441):1. https://doi.org/10.1126/science.aat3100

Chapter 2
Scientific Background

The following Chapter introduces three main topics of my work: ferroic order, nanomagnetism and metamaterials. First, the concept of ferroic order is presented. A symmetry-based classification is given together with a brief discussion of phase transitions, the emergence of a ferroic order parameter, spontaneous domain formation and the manipulation of an order parameter with a conjugate field. This part closes by unravelling ferrotoroidicity. Second, magnetic properties of sub-micrometre-sized objects made from a ferromagnetic material are discussed. Here, the formation of different kinds of spin structures is explained that serve as building blocks of nanomagnetic arrays. The suppression or—more important here—the support of long-range order in extended magnetostatic-coupled arrays is explained. The third part of this Chapter introduces metamaterials, a class of matter that is assembled on length scales comparable with the wavelength of radiation that interacts with it and that provides design-determined novel material properties and functionalities.

2.1 Ferroic Order

Ferroic order, mostly known for the exemplary case of ferromagnetism, is referred to as a classification of materials that exhibit a spontaneous macroscopic order and can be manipulated by the application of an appropriate field [1–5]. The ferroic state is connected to a phase transition at which a so-called *order parameter* spontaneously emerges. Following the Curie principle [6], this parameter formally describes the newly arised material characteristics and related symmetry properties. In particular, the violation of spatial point symmetries (e.g. space inversion) and/or temporal symmetries (i.e. time reversal), once a material undergoes a phase transition, allow for distinct physical phenomena in the ferroic phase (see Sect. 2.1.1). The concordant orientation of the order parameter (see Sect. 2.1.2) due to cooperative microscopic interactions is connected to the formation of a spatially extended and homogeneously ordered ground state. Ground-state degeneracy and competing con-

J. Lehmann, *Toroidal Order in Magnetic Metamaterials*, Springer Theses,
https://doi.org/10.1007/978-3-030-85495-9_2

tributions to the free energy of the system cause the spontaneous emergence of an ensemble of areas in which the order parameter is uniformly aligned. These ferroic domains (see Sect. 2.1.3) constitute equivalent realisations of the order, yet with different orientations of the order parameter. A key property of ferroic crystals is the ability to change the orientation of the order parameter by means of the application of a conjugate field (see Sect. 2.1.4). The flexibility to externally manipulate a material's physical observable and, in addition, potential cross-coupling between physical measures are key for their broad use as active material in transformers, capacitors, sensors, actuators, memory cells or the like. The following sections introduce essential aspects of the ferroic state from different perspectives.

2.1.1 Symmetry Considerations and Overview of Ferroic States

A crystalline material can be mathematically described as a periodic lattice in space whose constituents fulfil particular point symmetries that—as a whole—determine fundamental material characteristics. A coarse assignment with respect to the formally allowed properties according to the Neumann [7–9] or more general the Curie [6] principle can be made from the crystal's point-group symmetry, which is given by the combination of its rotation and inversion symmetries. As a consequence, tensors that describe physical properties of a material are invariant under the symmetry operations of its point group. For the discussion of ferroic materials, time-reversal symmetry is of equal importance as it allows for describing magnetically ordered materials [1, 7, 10–13]. Here, magnetic moments or spins are seen as semi-classical current loops that change their sense of rotation upon time reversal.
The general classification of symmetry in crystals is well established in terms of c- and i-type tensors [7, 14] (where i stands for invariant and c for change, with respect to time-reversal symmetry). For my brief descriptions of material properties in ferroics, I limit the discussion and differentiate only between two different sets of vectors corresponding to a particular arrangement of charges or spins: polar vectors and axial vectors of either i- or c-type [8, 10, 15–17]. While polar i-type vectors break space-inversion symmetry and can be associated with a charge displacement (an electric polarisation $\vec{P}$), axial c-type vectors break time-reversal symmetry and relate to the curl of a polar vector (e.g. an electric loop current semi-classically generating a magnetic moment $\vec{\mu}$). How both types of vectors behave differently under the application of spatial or temporal symmetry operations can be seen in Fig. 2.1.
Ferroic crystals with an electric or magnetic order require an appropriate description in terms of this symmetry concept. The complete set of point-symmetry operations (and time-reversal symmetry) that leaves the spin and charge structure unchanged is given by the magnetic point group. Since knowledge about the magnetic point group gives insight into allowed physical properties and the anisotropy associated to their descriptive tensors, it is useful to sort crystals with respect to their behaviour under characteristic symmetry operations [1, 18–20], see Table 2.1.

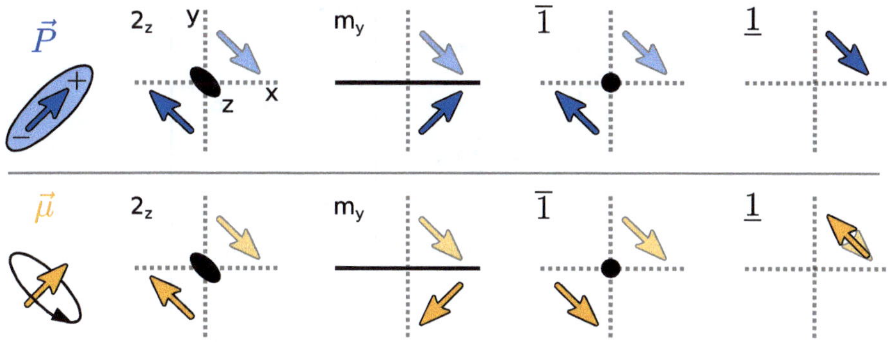

Fig. 2.1 Basic symmetry operations on polar and axial vectors. Polar vectors (upper line), as for example the electric polarisation $\vec{P}$ (blue arrows) due to charge displacements and axial vectors (lower line) such as magnetic moments $\vec{\mu}$ (orange arrows) behave differently under symmetry operations. The application of a two-fold rotation (2_z), a mirror plane (m_y), space inversion ($\bar{1}$) and time reversal ($\underline{1}$) onto both types of vectors is depicted. Pale arrows: initial orientation; Solid arrows: resulting orientation. Note that particular symmetry operations affect both the vector's position with respect to the origin as well as its direction

Table 2.1 Physical consequences of a material's behaviour under symmetry transformations. A mere look at the symmetry elements for a particular material (left-hand side) allows to sort them with respect to a fundamental physical character (right-hand side)

Symmetry character	Physical consequence
$\nexists$ Space-inversion symmetry ($\bar{1}$)	non-centrosymmetric
$\nexists$ Mirror symmetry ($\bar{N}$)	chiral
$\exists$ Directional axis	polar
$\nexists$ Time-reversal symmetry ($\underline{1}$)	magnetically ordered
$\nexists$ Time- and space-inversion symmetry ($\underline{1}$, $\bar{1}$)	lin. magnetoelectric [21–23]

Now, for a given set of symmetry violations or invariances, it is—a priori—possible to determine if a crystal can or can not host a particular ferroic order and, further, how it may respond upon the application of fields [7, 24–27]. Such a classification can be a first step in the search for crystals permitting a desired behaviour, see Fig. 2.2. If the ferroic phase is characterised by a single macroscopic moment fully describing its order and accounting for its symmetry, and if it is switchable by a single external field, it is called a *primary* ferroic state [1, 17, 18]. However, if a combination of two or more fields is required to switch the particular order, the state is referred to as being of *secondary, tertiary, ...* ferroic nature [2–4, 17, 28]. As displayed in Fig. 2.2, four basic classes of primary ferroic states are commonly distinguished with respect to their behaviour under space and time parity operations.

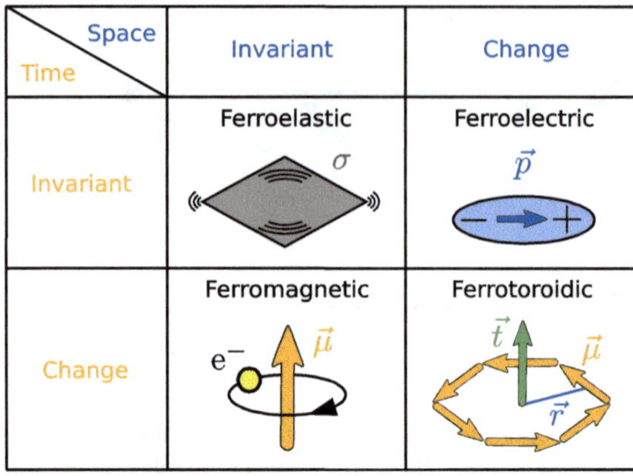

Fig. 2.2 Symmetry classification of four primary ferroic states. Categorised is the parity of different ferroic states upon time-reversal- and space-inversion-symmetry operation (Eigenvalue of -1: change; $+1$: invariant). The origin of magnetic moments $\vec{\mu}$ (orange arrows) is depicted as an electronic current, while the origin of electric dipoles $\vec{p}$ (blue arrow) is indicated as a displaced charge. Ferrotoroidic order is based on whirls of magnetic moments creating a toroidal moment $\vec{t}$ (green arrow). Figure modified from Ref. [29]

Ferroelastic materials [30, 31] exhibit a deformation of the unit cell and, thus, a spontaneous strain that can be switched by applying mechanical stress.[1] For this, a broken rotational symmetry and, hence, a change of point symmetry is required, while the state remains even under time-reversal and space-inversion symmetry.

Ferroelectric materials [35] exhibit a spontaneous macroscopic polarisation and thus, collectively aligned microscopic dipole moments that can be switched by an electric field. The electric dipole as a polar-vectorial property requires a non-centrosymmetric point group and, furthermore, at least one directed crystal axis implying pyroelectricity.

Ferromagnetic materials [36–39] exhibit a spontaneous collective ordering of magnetic dipole moments and, thus, a macroscopic magnetisation that can be switched by a magnetic field. Ordering of this axial-vectorial property results in the violation of time-reversal symmetry. Besides the widely known parallel alignment of magnetic moments in ferromagnets, more complex types of magnetic order with a macroscopic magnetisation exist, namely ferrimagnets (partially compensated alignments

---

[1] Since the spontaneous strain is described by a second-rank tensor instead of a vector, like for the other three primary ferroic states, it appears detached from them [17]. To circumvent this issue ferro-rotational order has been proposed recently as a replacement with a suitable symmetry and a vectorial order parameter [32–34].

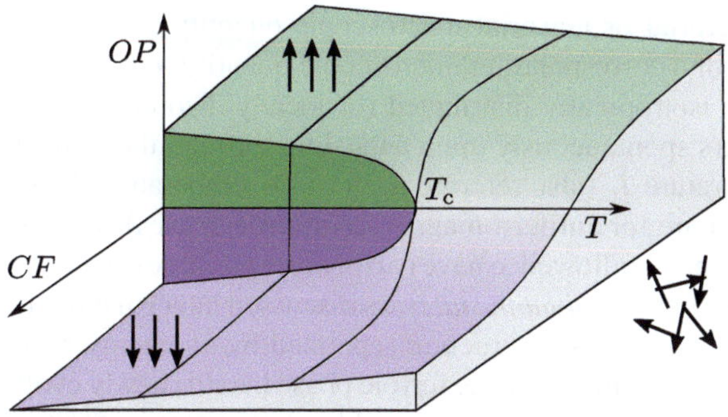

Fig. 2.3 Phase diagram of a ferroic system. Schematic temperature- (T) and conjugate-field-dependent *(CF)* behaviour of the order parameter $OP(CF, T)$ in equilibrium, see Sect. 2.1.4. Annealing at zero field results in a continuous phase transition with spontaneous symmetry breaking at the critical temperature T_c, accompanied by the emergence and growth of the *OP* with—here—two-fold degeneracy (up: green, down: violet). In the ferroic phase, the system undergoes a field-induced first-order phase transition with a discontinuity at $CF = 0$

of magnetic moments) or certain types of non-collinear magnets (e.g. uncompensated helical, sinusoidal or cycloidal arrangements of magnetic moments).

Ferrotoroidic materials [16, 40–43] are proposed to represent the fourth type of primary ferroic state that exhibits a spontaneous toroidisation via a collective alignment of magnetic whirls that are switchable via e.g. locally curled magnetic fields. The emergence of a toroidisation, a polar c-type vector, leads to the simultaneous violation of space-inversion and time-reversal symmetry. To comply with this symmetry condition, certain arrangements of magnetic moments such as a magnetic monopole or by a magnetic quadrupole can be thought of, too [44, 45]. However, as shown later, in particular the ferrotoroidic state is a preferred candidate since a conjugate field has been identified and microscopic interactions to stabilise this exotic magnetic state are known to exist in some materials.

2.1.2 Ferroic Phase Transition and Order Parameter

External parameters such as temperature, pressure or applied fields determine the particular phase in which a material is situated. The transition from a high-symmetry so called *parent* or *prototypical* phase into a low-symmetry ferroic phase is accompanied by the spontaneous loss of point or temporal symmetry operations [1, 46]. In 1937, Landau [47] introduced the concept of a thermodynamic quantity called the *order parameter* (*OP*) in a mean-field theory about continuous phase transitions which typically relate to ordering processes in crystals [17]. Implied by Landau theory, a symmetry consideration of the parent and the ferroic phase gives access to the point symmetry of the suitably defined order parameter itself. The order parameter

as a scalar, vectorial or tensorial macroscopic quantity is a sensitive measure for the correlation of a corresponding microscopic quantity (e.g., the density of aligned dipoles) that is isotropically distributed (or locally disordered) in the prototypical phase and orders spontaneously upon annealing through the critical temperature or ordering temperature T_c (also referred to as Curie temperature for most ferroics and as Néel temperature for antiferromagnets/antiferroelectrics), see Fig. 2.3.
The order parameter is allowed to have two or more possible orientations (corresponding to different so-called *domain states*, as discussed later) that relate to local energy minima in the ferroic crystal structure separated by an energy barrier. As a consequence, the phase transition into the ferroic phase heralds newly evolved microscopic as well as macroscopic properties allowed by the symmetry of the order parameter (or equivalently by the symmetry that has been broken during the transition). These properties, e.g., the magnetisation $\vec{M}$ in ferromagnetic or the polarisation $\vec{P}$ in ferroelectric materials are measurable variables of the ferroic crystal and quantify the response of the system upon the application of a conjugate field CF (a magnetic field $\vec{H}$ for ferromagnetic or an electric field $\vec{E}$ for ferroelectric materials, as discussed in the following section). Here, the material-dependent susceptibility χ quantifies physical interactions as it relates the cause (conjugate field, CF) with the response (order parameter, OP) asquantifies physical interactions as

$$OP = \chi \; CF \quad \longleftrightarrow \quad \chi = \frac{\partial}{\partial CF} \, OP, \tag{2.1}$$

For continuous phase transitions, the susceptibility χ diverges at the critical temperature T_c. Far above T_c, this behaviour is described by the Curie-Weiss law, which is based on a mean-field approximation. In close vicinity to T_c, however, critical behaviour due to fluctuations requires a more accurate description based on the introduction of so-called *critical exponents* γ and β.

$$\begin{aligned} \chi &\propto (T - T_c)^{-\gamma} \text{ for } T > T_c \\ OP &\propto (T - T_c)^{\beta} \text{ for } T < T_c \end{aligned}, \tag{2.2}$$

with $\gamma = 1$ and $\beta = 0.5$ in the mean-field approximation.[2] The order parameter is zero at T_c, however. The correlation length as the characteristic length scale over which fluctuations of microscopic degrees of freedom are correlated, peaks just like the susceptibility at T_c implying a macroscopic correlation. Experimentally, susceptibilities related to the order show a strong enhancement at continuous phase transitions (and a mere discontinuity for first-order phase transitions) which attracts great interest from both fundamental and technological perspectives.
The phase transition can be addressed from a statistical-mechanics point of view. By assuming the conservation of energy (E) and the minimisation of free energy ($F = E - TS$, where S is the entropy and T the temperature), it can be shown that

[2] Generally, these exponents are expected to be of universal character and, as such, only dependent on global details of the investigated system like its physical dimension, or the range of the interaction.

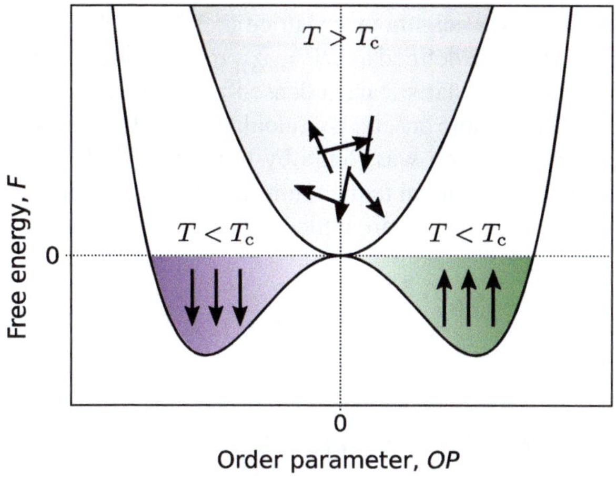

Fig. 2.4 Free-energy diagram above and below the critical temperature. Shown are two curves representing the free energy F as a function of the order parameter OP for the high- and the low-temperature phase. Above the phase transition temperature T_c, the average order parameter is zero. Below T_c, the reformation of the energy landscape favours a non-zero order parameter OP with orientational degeneracy (up: green, down: violet)

the probability P_i, for the occurrence of any microscopic state i with energy (E_i) is given by the Boltzmann distribution,

$$P_i = \frac{g_i}{Z} \exp\left(\frac{-E_i}{k_B T}\right), \tag{2.3}$$

where g_i is the degeneracy of the considered state, $k_B T$ is the thermal energy, with Boltzmann constant k_B and partition function Z for normalisation. A variation of the system's temperature changes its free energy such that a particular macroscopic state can be energetically favoured below a critical temperature (T_c), see Fig. 2.4. The system will therefore undergo a phase transition and adapt the configuration and energy landscape present in the new phase.

Various models exist to exploit the nature of phase transitions, with the so-called *Ising model* being the most studied one [48–51]. Its relevant microscopic variables are one-component vectors σ_i, typically spins, with a binary degree of freedom that just allows a parallel or antiparallel alignment with respect to a particular axis. These two-state objects occupy sites on a lattice and are connected to its direct neighbours via an interaction parameter J. The Ising Hamiltonian H_{Ising} that describes the interaction is written as

$$H_{\text{Ising}} = -J \sum_{i \neq j} \sigma_i \sigma_j, \tag{2.4}$$

where the sum is taken over neighbouring lattice sites. The value of a global Ising-like order parameter can now be defined as $OP = \sum_i \sigma_i$. The two-dimensional variant of the Ising model is one of the statistical condensed-matter models that displays a phase transition from a rather simple ansatz. By calculating the free energy of an Ising-like system, an analytic expression was found by Lars Onsager [52] that calculates an exact relation between the critical temperature T_c and anisotropic nearest-neighbour interaction parameters J_i on a square lattice for zero applied field:

$$\sinh\left(\frac{2J_1}{k_B T_c}\right) \sinh\left(\frac{2J_2}{k_B T_c}\right) \equiv 1. \tag{2.5}$$

2.1.3 Spontaneous Domain Formation

The spontaneous point- and/or time-symmetry-breaking decomposition of the parent phase into a characteristic number of differently oriented regions with a homogeneously aligned order parameter of the ferroic phase, called *domain states*, is a hallmark property of the ferroic phase.

Ferroic domains are macroscopically extended regions with a uniform direction of the order parameter which spontaneously form when a material passes a ferroic phase transition [1, 4]. One particular domain state stands for one particular realisation of the ferroic phase while the number of possible domain states is determined by the lost symmetry operations during the phase transition. The physical reason for domain formation can be of different origins. Formally, it is the balance of different energy contributions that determines the actual realisation of a domain structure. In ferromagnets, as hitherto the most relevant class of ferroics, domains exist in the ground state in thermal equilibrium since, in general, a multi-domain state yields lower internal energy than a single-domain state, see details in Sect. 2.2. On the one hand, a ferromagnet gains magnetic-exchange energy $E_{ex} \propto -\cos(\theta_{ij})$ with non-parallel aligned magnetic moments (with θ_{ij} being the angle between neighbouring moments), which would exist inside a domain wall, see below. Therefore the exchange interaction favours a single-domain state across the whole crystal. Since the exchange interaction in this example relates to the energy cost for a domain wall, it macroscopically scales with an area ($E_{ex} \propto r^2$, with r parametrising the domain size). On the other hand, a ferromagnet gains magnetostatic energy from its macroscopically extending stray field or demagnetising field. Therefore, magnetostatic energy, E_{ms}, favours to split the material into a large number of randomly oriented domains, minimising the generated field [53]. The magnetostatic energy scales with the volume of a domain or of a crystal ($E_{ms} \propto r^3$ with r accounting for the size of the domain or the crystal). The simplified situation of a mere interplay of these two competing

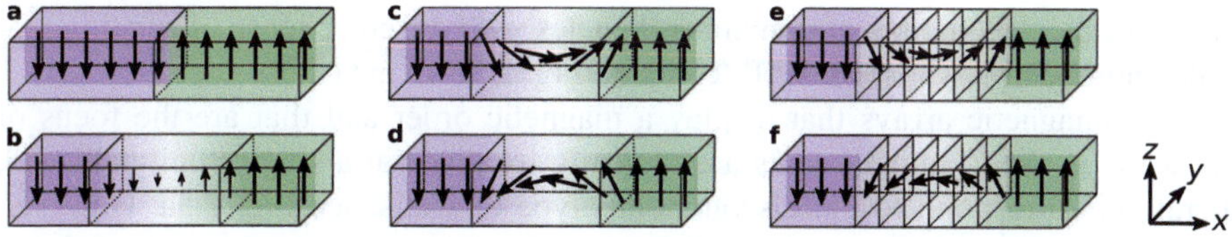

Fig. 2.5 Variants of ferroic domain walls. Different types of boundaries between $-OP_z$ (left, violet) and $+OP_z$ (right, green) are shown. **a** A localised and discontinuous transition at which the order parameter reverses the sign in adjacent unit cells. **b** A gradual decrease and subsequent increase of the amplitude of the order parameter. **c–d** Clockwise and counter-clockwise rotations of the order-parameter direction in the xz-plane perpendicular to the domain wall, so-called *Néel walls*. **e–f** Clockwise and counter-clockwise rotations of the order-parameter direction in the yz-plane parallel to the domain wall, so-called *Bloch walls*

terms[3] already results in a multi-domain state in thermal equilibrium. For magnetic types of long-range order with vanishing net magnetisation, as for instance in antiferromagnetic or ferrotoroidic materials, the situation is entirely different. Here, the domain-wall energy exists as well, but the magnetostatic energy, or a corresponding analogue measure for the particular ferroic state as its counterpart, is absent. Thus, in thermal equilibrium and at finite temperature, mainly the thermal contribution to the free energy, following Eq. 2.3, allows for the formation of a multi-domain state [57]. For this reason, a non-equilibrium process, e.g. a temperature quench, is necessary for the spontaneous formation of domains.

As mentioned above, different domain states are separated from one another by a domain wall, which can be seen as a natural interface between the two states [58–62]. Within the domain wall, the direction of the order parameter has to make a transition from one orientation to the other. Several possibilities exist for realising this, see Fig. 2.5.

The interplay of energy terms drive the emergence of a domain wall of a specific kind and determines its width. In the case of ferromagnets, the competition between the anisotropy energy that favours an order-parameter alignment along a particular crystalline direction, and the exchange-interaction energy that favours an (anti)parallel orientation of neighbouring moments, determine the width of a domain wall. The type of the wall can be determined typically by considering the domain wall as an alone standing entity. Thus, by comparing the internal energy of different possible realisations of a domain wall, a statement about its preferential configuration can be made. While it is often the sample geometry and anisotropy energy that determine the general type of the wall, its microscopic details, e.g., the sense of rotation require a more precise analysis. Here, peculiarities in the underlying exchange interaction that result from e.g. spin-orbit interaction associated with for instance the Dzyaloshinskii-Moriya interaction (DMI [63, 64]) in non-centrosymmetric systems

[3] For a more detailed study, energy contributions from the magnetic anisotropy (see Sect. 2.2), magnetostriction, strain, crystal defects and grains or annealing procedures have to be taken into account [53–56].

may split the degeneracy of domain-wall states with different sense of rotation (so-called homochiral walls) [65–70] (compare Fig. 2.5c–d and e–f).
For nanomagnetic arrays that display a magnetic order and that are the focus of the work at hand, domain walls are typically localised at a single 'unit-cell' and, therefore, correspond to the discontinuous type of walls, see Fig. 2.5a. However, despite being discontinuous, the walls in those systems can indeed be associated to different configurations of a particular (but rather discrete) sense of rotation as shown Fig. 2.5c–d and e–f.

2.1.4 Conjugate-Field Poling

As introduced above, a key property of the ferroic state is the possibility for a hysteretic reversal of the order-parameter direction via the application of an external conjugate field. The reversal of an order parameter, called switching (often applied to ferromagnets) or poling (often applied to ferroelectrics), costs energy since it requires the nucleation of oppositely oriented domains and their expansion throughout the crystal. Note that due to the bias caused by the field, a reversal is accompanied by a decrease of the total energy in the presence of the field. The applied field manipulates the internal energy landscape and lifts the zero-field degeneracy of the ground state. In a ferromagnetic material, for example, the conjugate field is a magnetic field $\vec{H}$ reflecting the symmetry of the order parameter, the magnetisation $\vec{M}$. Here, application of a magnetic field results in a Zeeman interaction that introduces an energy imbalance between magnetic moments $\vec{\mu}$ of different domain states and results in a modification of the internal energy by $E_Z \propto -\vec{\mu}\vec{H}$, termed the *Zeeman energy*. Domain states with a magnetisation parallel to the external-field direction reduce their internal energy by the Zeeman energy and thus tend to grow, whereas differently oriented domains increase their internal energy accordingly and become unfavoured and, hence, shrink.
Macroscopic information about the reversal process is captured by the hysteresis curve that describes the variation of the average order parameter as a function of the present state and the applied field, see Fig. 2.6. Starting with a multi-domain state in thermal equilibrium (Fig. 2.6a), the application of a conjugate field in positive direction favours one state over the other and aligns the order parameter with respect to the field direction gradually into saturation (Fig. 2.6b). In a ferromagnet, for example, the spins are now aligned parallel to the external field, denoting a single-domain ferromagnetic state; the material is fully magnetised. From here, a monotonic decrease of the external field back to zero is accompanied by a residual net value of the order parameter (Fig. 2.6c), called the *remanence*, R (remanence magnetisation for the case of ferromagnets). A further increase of the external field in the negative direction up to the amplitude of $-CF_C$, the so-called *coercive field*, generates equally populated domain states (Fig. 2.6d), leading to a vanishing average order parameter. Finally, negative saturation is reached but with an oppositely oriented reversed order parameter (Fig. 2.6e).

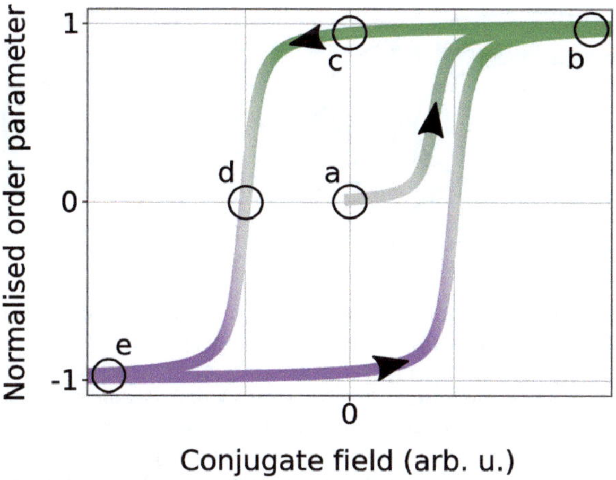

Fig. 2.6 Typical ferroic hysteresis curve. Shown is the normalised order-parameter average as a function of the conjugate-field strength. A conjugate field imbalances the free-energy landscape (Fig. 2.4) of a virgin multi-domain state (a), lifts the degeneracy and induces domain nucleations and domain-wall movements that favour one state over the other (violet and green). The system eventually reaches saturation with a fully aligned order parameter (b). A subsequent decrease of the field to zero yields a non-zero remanence of the order-parameter average (c). An increasingly negative field enlarges oppositely aligned domains until the coverage of both domain-state orientations is equal at the coercive field (d). Finally, a complete reversal of the order parameter is reached at the negative saturation field (e). The hysteresis curve is modelled phenomenologically with two Langevin functions given as $L(CF) = \coth(CF \pm CF_C) - (CF \pm CF_C)^{-1}$, with CF being the conjugate field and CF_C defining the opening of the hysteresis curve and, thus, the particular field strength required for switching

2.1.5 *The Toroidal Moment and Ferrotoroidicity*

Ferrotoroidicity as the fourth type of primary ferroic state is characterised by a long-range order of magnetic moments that spontaneously form hysteretically switchable vortex states, so-called toroidal moments, of a defined sense of circulation within a unit cell [42]. The toroidal moment as such has been reviewed comprehensively in scientific literature [40, 41, 71–74], while ferrotoroidicity [42, 43, 75–78] is still poorly understood and has recently been described as the "most elusive category of primary ferroic orders" [78]. In the following sections, the toroidal moment as an isolated entity is introduced from a classical and a quantum-mechanical perspective. Further, the collective behaviour of toroidal moments in a crystalline solid [42], its description with respect to the definition of ferroic states [16] and its connection to the linear magnetoelectric effect [43] are discussed. The section closes with an overview over ferrotoroidic domains and an analysis of demonstrated and potential ferroic poling schemes.

The Toroidal Moment

Microscopically, the toroidal moment is related to a magnetic flux-closure state that most-generally originates from a term in the multipole expansion [22] of the

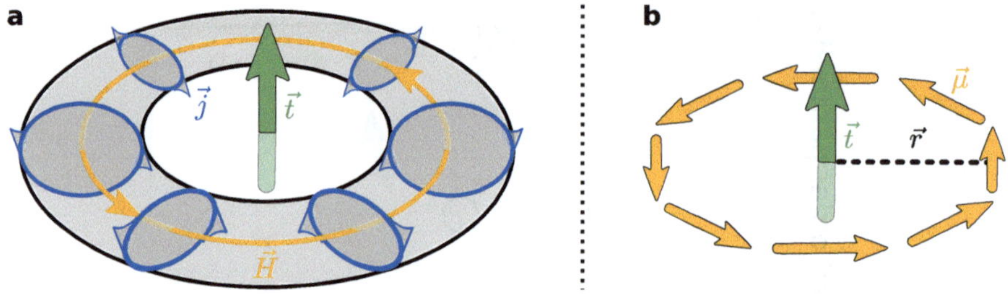

Fig. 2.7 Representations of a toroidal moment. **a** In classical electrodynamics, the toroidal moment $\vec{t}$ originates from a term in the multipole expansion of the electromagnetic vector potential and can be depicted by a poloidal current distribution $\vec{j}$ on the surface of a torus, generating a magnetic field $\vec{H}$ that bends into a circle. Note that this picture, despite being didactically helpful, does not represent well the actual situation in the context of solid-state physics as discussed within this book. **b** In quantum mechanics, the toroidal moment originates from a planar vortex-like arrangement of magnetic moments $\vec{\mu}$ (with spin or orbital contributions [81, 82]) displaced from their shared origin by $\vec{r}$

electromagnetic vector potential of an arbitrary charge and current distribution [41, 43, 79] that is not invariant under space-inversion and time-reversal symmetry. In this picture, a toroidal moment $\vec{t}$ is represented by a current density $\vec{j}$ on the surface of a torus that flows uniformly in poloidal direction (or recursively defined as $\vec{j} \propto \vec{\nabla} \times (\vec{\nabla} \times \vec{t})$) [80]. Such a current distribution induces a curled magnetic field $\vec{H}$ in the volume of the torus along the toroidal direction, see Fig. 2.7a, which constitutes the electrodynamic source of the toroidal moment.

It is important to note that this definition of magneto-toroidal moments and magneto-toroidal order differs significantly from closely related concepts such as the electro-toroidal moment or a dynamic or transient toroidal moment. The symmetry of an electro-toroidal moment, a compensated vortex-like arrangement of electric dipoles, is symmetric under both space-inversion and time reversal operation [41, 83, 84] and, thus, constitutes an entirely different class of matter. Dynamic toroidal moments are electromagnetic excitations and resonances, observed in suitably designed metamaterials, that require a description beyond the typical magnetic- or electric-dipole picture [72, 74, 85]. Such magneto-toroidal high-frequency modes rely on current distributions that form simplified variants of the construction introduced in Fig. 2.7a. However, these excitations have an AC character, and therefore change sign every half cycle of the exciting AC field. Consequently, these modes do not allow the formation of stable and static toroidal systems. In the framework of this book, toroidal order is related to magnetically ordered systems with ferroic character, satisfying the ferroic hallmark properties of spontaneous domain formation, a macroscopic order parameter and hysteretic poling in a conjugate field.

Toroidal Order in Crystalline Solids—Ferrotoroidicity

In solid-state physics, the microscopic origin of a toroidal moment $\vec{t}$ is a spin density $\vec{\mu}(\vec{r})$ with a whirl-like magnetically compensated arrangement at the unit-cell level, see Fig. 2.7b, and was first studied[4] in the 1980s by Soviet scientists [40, 41, 71, 75, 80, 86]. The toroidal moment is mathematically defined as an integration over a unit cell (uc) [42],

$$\vec{t} = \frac{1}{2} \int_{\mathrm{uc}} \vec{r} \times \vec{\mu}(\vec{r}) \, d^3r. \tag{2.6}$$

Hence, the toroidal-moment vector is oriented perpendicular to the plane in which the magnetisation whirls and transforms as a polar c-type vector [75, 80, 86], see Sect. 2.1.1.

Ferrotoroidicity is referred to as a collective spontaneous alignment of magnetic vortex states with the same sense of vorticity. It yields a macroscopic toroidisation $\vec{T}$ as the corresponding order parameter [43, 75]. For a periodic crystal containing discrete magnetic moments $\vec{\mu}_i = \vec{\mu}_0 \delta(\vec{r} - \vec{r}_i)$, the toroidisation can be expressed as a density and, as a result, does not depend on details of a finite-system such as the number of unit cells N [42];

$$\vec{T} = \frac{1}{2NV} \sum_{\mathrm{uc}}^{N} \sum_{i} \vec{r}_i \times \vec{\mu}_i = \frac{1}{2V} \sum_{i} \vec{r}_i \times \vec{\mu}_i. \tag{2.7}$$

Here, V is the unit-cell volume and $\vec{r}_i$ is the position of the ith magnetic moment $\vec{\mu}_i$, while the summations are performed over all moments in a unit cell and over all considered unit cells, respectively. However, the definition of local toroidal moments in a crystal suffers—in a similar way as the electric-dipole moment in ferroelectrics—from a multivaluedness that results from periodic boundary conditions and the freedom of the particular choice of the unit cell, or the magnetic basis vectors [42, 87]. Mathematically, this consideration can be described by a re-defined spatial coordinate: $\vec{r} \mapsto \vec{r}' = \vec{r} + \vec{R}$, with $\vec{R}$ being a primitive lattice vector (capturing the periodic boundary conditions) or a non-primitive translation vector (capturing the freedom-of-origin dependence). The toroidisation $\vec{T}$, thus, shows a multivaluedness with an increment of $\Delta\vec{T} = \frac{1}{2}\sum_i \vec{R} \times \vec{\mu}_i$. Due to that multivaluedness, only changes (during a phase transition or during an annealing) or differences (between different regions of the sample) in the toroidisation are well-defined measures for any consistent choice of the unit cell [42, 43] in Eq. 2.7.

2.1.6 *Properties of the Ferrotoroidic State*

Ferrotoroidicity requires the spontaneous occurrence of two or more distinct domain states that can be described by the orientation of a primary order parameter as introduced in Sect. 2.1.1. In case of toroidal order, this scenario is satisfied by uniform

[4] Note that the so-called anapole moment, first mentioned already in 1958 [79] is a closely related term for the non-radiative mode of an electromagnetic excitation with the same symmetry.

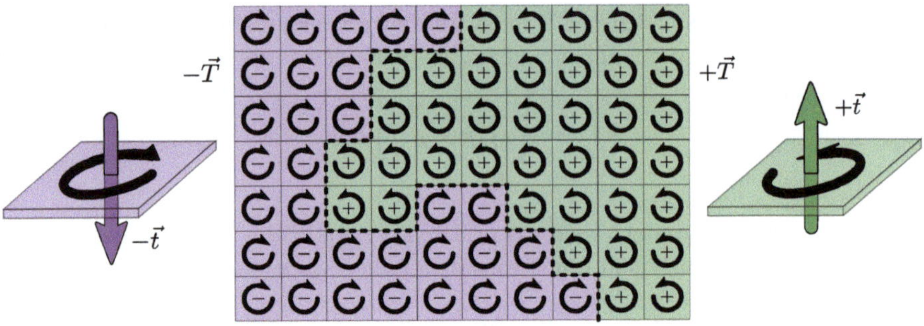

Fig. 2.8 Ferrotoroidic domain structure. Sketched are 10 × 7 unit cells, each containing a toroidal moment ($-\vec{t}$, $+\vec{t}$) from a vortex-like arrangement of magnetic moments (black circular arrows). One domain wall (dashed line) crosses the system and separates a $-\vec{T}$ domain (violet) from a $+\vec{T}$ domain (green)

regions of clockwise ($-\vec{T}$) or counter-clockwise ($+\vec{T}$) arranged magnetic moments, as depicted in Fig. 2.8. Imaging these ferrotoroidic domains is much more challenging as compared with the observation of, e.g., ferromagnetic or ferroelectric domains. While in ferromagnets, the magnetisation as the order parameter provides a measurable external stray field that can be detected by different experimental techniques (see, e.g., Chap. 3.2), in magnetically compensated ferrotoroidic materials, no external magnetic field is present. Furthermore, the coupling of measurable properties to the toroidisation in a material is of rather subtle nature. Therefore, a suitable measurement technique possibly requires an indirect sensing of the spin-driven change in crystal symmetry or a coupling to the order parameter via optical second-harmonic generation (SHG) microscopy [88–92], a measurement of the magnetic order via spherical neutron polarimetry [93–96] and x-ray gyrotropy or dichroism [82, 97, 98] or magnetoelectric measurements [43, 99].
Crystals of $LiCoPO_4$ exhibit long-range toroidal order in their low temperature phase below a temperature of 21.8 K [29, 100]. A weak toroidisation emerges in the ferrotoroidic phase due to the simultaneous displacement of two pairs of spins inside a unit cell: Two spins displace outwards and increase their toroidal moment and two spins displace inwards and decrease their oppositely aligned toroidal moment. This arrangement can, in analogy to partly compensated ferrimagnetic materials, be referred to as a ferritoroidic state. Toroidal domains and their coexistence with antiferromagnetic domains have been imaged in $LiCoPO_4$ using SHG as symmetry sensitive probing technique [29]. Here, the non-linear interaction of the crystal with an incoming light wave leads to an optical signal at twice the frequency (i.e. *second harmonic*). The second-harmonic signal transfers information about, among others, symmetry violations due to the macroscopic order in the crystal. This information is encoded in the light polarisation and phase and can thus be used for the imaging of domain configurations [92].

Symmetry Analysis and Candidate Systems

A symmetry analysis allows to identify crystals that are suitable for hosting ferrotoroidic order. It can be found in 31 out of 122 magnetic point groups that allow for the existence of a polar c-type (also called axio-polar) vector [16, 71, 101, 102]:

$$\left\{ \begin{array}{l} 1, \bar{1}, 2, \underline{2}, \bar{\underline{2}}, 2/\underline{m}, \underline{2}/m, 3, \bar{\underline{3}}, 3m, \bar{3}\underline{m}, 3\underline{2}, 4, \bar{\underline{4}}, 4/\underline{m}, 4mm, 4/\underline{m}mm, \bar{4}m\underline{2}, \\ 4\underline{22}, 6, \bar{\underline{6}}, 6/\underline{m}, 6mm, 6/\underline{m}mm, \bar{\underline{6}}m\underline{2}, 6\underline{22}, m, mm2, m\underline{m2}, \underline{m}mm, 2\underline{22} \end{array} \right\},$$

where ($\underline{N}$) and ($\overline{N}$) indicates the particular point-symmetry operation (N) combined with time reversal and space inversion, respectively. Several compounds are under discussion or have already been identified as candidates for ferrotoroidic crystals: Li(Co, Fe, Ni)PO_4 [29, 100, 103–107], $Ga_{2-x}Fe_xO_3$ [108–114], $LiFeSi_2O_6$ [99, 115, 116], $BiFeO_3$ [117–119], (Co, Fe, Ni)$_3B_7O_{13}$(Cl, Br, I) [120, 121], $MnPS_3$ [122], Cr_2O_3 in magnetic fields [123], (Tb, Dy, Ho)-molecular clusters [124–132], and others [78].

Relation to Antiferromagnetism, Primary Ferroic Nature and Magnetoelectricity

Because of the compensated arrangement of magnetic moments in a toroidally ordered crystal, the ferrotoroidic state suggests itself that it is a subclass of antiferromagnetism. Therefore, its primary ferroic character has been questioned. Furthermore, the character of the ferrotoroidic state with breaking time-reversal and space-inversion symmetry implies an intrinsic affiliation to magnetoelectric multiferroics [133–137] or to a ferromagnetoelectricstate [1, 2, 16, 17, 40] (of secondary ferroic nature). This overlap and ambiguity with different ferroic classes challenges its separation as a stand alone primary ferroic order [138]. Yet, the ferrotoroidic state underlines its fundamental importance, because it provides an inherent linear magnetoelectric coupling [15, 43, 71, 76, 80] and possibly allows for advanced spintronic applications [139–144].

Relation to antiferromagnetism: The difficulty in distinguishing ferrotoroidic order from antiferromagnetic order is mainly due to the entwining of the microscopic definition of the underlying spin structures [16, 17]. On a fundamental level, the prototypical antiferromagnetic state is microscopically defined by the sign of the exchange integral that results in a 180° angle between neighbouring spins. Macroscopically it requires the existence of a double hysteresis upon the application of a magnetic field. Beyond this overall similarity, the ferrotoroidic state allows for certain features due to its polar c-type symmetry, but not readily the other way around. The order parameter in antiferromagnetic materials is often defined by an abstract notation that is, in general, not related to macroscopic physical properties linked one-to-one with the domain state. Contrastingly, the ferrotoroidic order parameter necessitates such a physical distinction between different toroidal domain states [26, 71, 76], exhibits a conjugate field and constitutes a proper manifestation of the state in which the order parameter has been identified to drive the ferroic phase transition [99]. Hence, it fulfils the definition of a primary ferroic order as discussed below. An example of a macroscopic consequence of the ferrotoroidic state is a directional electrical response function can be measured at the ferrotoroidic phase transition. It originates from the emergence of a net toroidization [138], resulting from the displacement current that is connected to the spontaneous alignment of toroidal moments. Additional physical phenomena accompanied with the ferrotoroidic phase transition are, e.g., a photogalvanic effect [80], nonreciprocal photon-momentum spectra [71], anomalous diamagnetism [75, 80], and a divergent linear magnetoelectric susceptibility [43]. These features are mutual properties of the ferrotoroidic phase and not general characteristics of the antiferromagnetic state.

Primary ferroic nature: Ferrotoroidicity is considered to be a primary ferroic state [17, 43] due to its spontaneously occurring singular order parameter and the existence of a solitary rather than a composed conjugate field, see the next section. A direct hint for the type of ferroic state would be knowledge about the leading interaction promoting the phase transition from the para- to the ferrotoroidic phase. In this context it has been shown that the transition to the ferrotoroidic phase in $LiFeSi_2O_6$ is very likely to be driven by a toroidal (vector-product-like) coupling and not by a scalar-product like coupling [99]. Further evidence is given by the toroidal-symmetry-connected observation of ferrotoroidic domains [29, 100]. Regardless of the final classification of the ferrotoroidic state, its unique transformation symmetries and exotic physical properties separates it from the established ferroic states. However, an explicit and experimentally clear separation is—to the best of my understanding—yet to be established.

Linear magnetoelectric effect: Ferrotoroidic order is inherently connected to the linear magnetoelectric effect [41, 43, 71, 76, 80], which describes a cross-coupling of magnetic and electric material responses [21, 22, 102, 145–148]. This behaviour can be derived from a free-energy expression and measured [120, 137] by the induction of a magnetisation ($\vec{M}$) as a response to an electric field ($\vec{E}$) or an induced polarisation ($\vec{P}$) as a response to a magnetic field ($\vec{H}$) via the linear magnetoelectric tensor α_{ij} as [23, 43]

$$P_i = \epsilon_0 \chi_{ij}^{\mathrm{e}} E_j + \alpha_{ij} H_j \quad ; \quad M_i = \mu_0 \chi_{ij}^{\mathrm{m}} H_j + \alpha_{ij} E_j, \tag{2.8}$$

with $\chi_{ij}^{\mathrm{e,m}}$ being the electric and magnetic susceptibilities, and ϵ_0 and μ_0 being the permittivity and permeability of vacuum, respectively. Typical materials that show large linear magnetoelectric effects are naturally found among the magnetoelectric multiferroics—materials that display ferroelectric and (anti)ferromagnetic order in the same phase. The violation of time-reversal and space-inversion symmetry in these materials, see Sect. 2.1.1, which eventually have the same origin is likely to be the cause for large linear magnetoelectric coupling coefficients [133, 134, 149–155]. The particular symmetry of the magnetoelectric interaction tensor α is reflected by the symmetry of the underlying magnetic point group of the material. Since the symmetry groups that allow ferrotoroidicity are identical with the ones that comprise off-diagonal antisymmetric components in α ($\alpha_{ij} = -\alpha_{ji}$), a detection of these

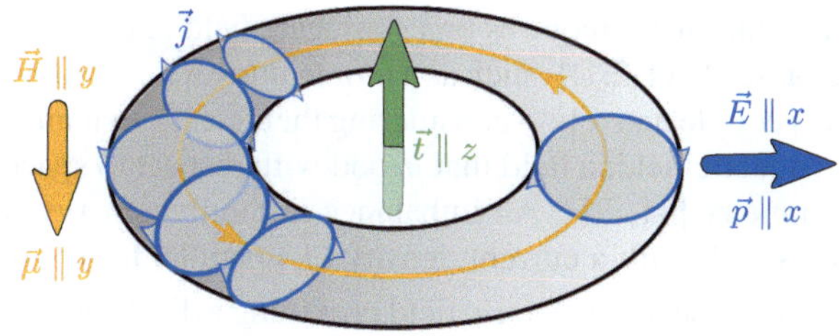

Fig. 2.9 Linear magnetoelectric nature of the toroidal moment. Depicted is the response of a toroidal moment $\vec{t} = t_0\vec{e}_z$, generated by poloidal currents on the surface of a torus, see Fig. 2.7, on the application of electric (blue arrow, $\vec{E} = E_0\vec{e}_x$) or magnetic (orange arrow, $\vec{H} = H_0\vec{e}_y$) fields. For example, a magnetic field in the plane of the torus (here along y) results in a Lorentz force that displaces the current loops (blue circular arrows) perpendicular to the field direction to lower the system's energy. The displacement of current is accompanied by the accumulating of charge. This charge imbalance causes the magnetoelectrically induced polarisation $\vec{p} \propto \vec{t} \times \vec{H}$. Figure adapted from Ref. [75]

magnetoelectric tensor components is desired and serves as an indicator[5] for ferrotoroidic candidate materials [42, 75, 78, 137]. By considering the classical picture of a current distribution on the surface of a torus as origin of the toroidal moment, the symmetry of the linear magnetoelectric effect upon application of a magnetic field becomes palpable, see Fig. 2.9. The excitation field (either $\vec{E}$ or $\vec{H}$) results in a displacement of charge currents perpendicular to the toroidal moment ($\pm\vec{t}$). The accumulated and thus spatially imbalanced charges now cause the generation of a magnetic, respectively, an electric dipole moment ($\vec{\mu}$ or $\vec{p}$) orthogonal to both. Formally, this interaction is described by a vector product of the toroidal moment $\vec{t}$ and excitation field (for instance, the magnetic field $\vec{H}$) as $p_i = \epsilon_{ijk} t_j H_k$ with ϵ_{ijk} being the Levi-Civita symbol [42].

Hysteretic Toroidal Poling and Nature of the Conjugate Field

As already mentioned, ferrotoroidic crystals do not yield a macroscopic electric or magnetic field that can be sensed [41]. Although in the classical picture of an ideal toroidal solenoid the entire electromagnetic field is concentrated inside the torus, a microscopic interaction has to exist in ferrotoroidic materials that connects neighbouring toroidal moments and allows for a collective behaviour. Indeed, toroidal moments interact for instance with external fields, as elaborated above for the case of the linear magnetoelectric effect.

Ferrotoroidicity as a primary ferroic order indeed implies the ability to hysteretically reverse the macroscopic toroidisation by the application of a conjugate field $\vec{S}$. The

[5] Confusion arises, in this context, due to an assumed identity of a toroidally ordered single-phase ferroic state and a multiferroic state that exhibits a magnetisation perpendicularly aligned to an electric polarisation [156–158]. The order in these magnetoelectric multiferroics, despite its potentially equivalent macroscopic symmetry, is not in agreement with the classification of the toroidal order as presented here [77], since it is neither magnetically compensated nor requires the same conjugate field for switching [42, 43].

ferrotoroidic order parameter requires this conjugate field to transform likewise as a polar c-type vector, see Sect. 2.1.1, such as a momentum or a current vector [16, 159]. Toroidal moments couple natively—considering their electrodynamic source term—to the curl of a magnetic field; a field that is odd with respect to space-inversion and time-reversal symmetry [41, 75]. An imbalance in the free energy, see Sect. 2.1.4, can be achieved as well with a current density ($\vec{j}_z$) parallel to the toroidal moment $\vec{t} = t_0\vec{e}_z$, with a time-dependent electric field providing a displacement current ($\frac{d\vec{E}_z}{dt}$), an in-plane magnetic field gradient ($\nabla H_{\perp z}$), or a product of perpendicularly aligned electric and magnetic fields ($\vec{E}_{\perp z} \times \vec{H}_{\perp z}$) [16, 17, 41, 71, 75, 160]. In particular, the last variant, addressing the linear magnetoelectric nature of the toroidal moment [16], appears experimentally feasible.

Toroidal poling has been successfully performed on $LiCoPO_4$, using SHG for imaging changes in the toroidal domain pattern and, thus, for recording the macroscopic toroidisation throughout the reversal process [100]. Here, perpendicular electric and magnetic fields have been applied for a magnetoelectric coupling to the toroidal field $\vec{E} \times \vec{H} \mapsto \vec{S}$ [41, 42, 102]. In addition, by probing a magnetic-field-induced pyroelectric current in $LiFeSe_2O_6$, the sign of the magnetoelectric tensor and, thus, the sign of the toroidisation has been reversed by means of an applied cross product of magnetic and electric fields [99]. Ever since, a direct generation of a conjugate toroidal field that is not composed of two separate fields remains an open challenge.

2.2 Magnetism in Confined Geometries

The systems that I studied during my work consist of a ferromagnetic alloy that is structured on the nanometre scale. In general, magnetism and magnetic order in solid-state physics are a quantum-mechanical consequence of microscopic couplings between atomic magnetic moments originating from the spin- or orbital angular momentum of bound or delocalised electrons. The spatial confinement of these interacting magnetic moments in nanostructures results in a broad range of magnetic textures whose realisation depends on the material of choice, the size and shape of the structures, their boundaries and surrounding magnetic fields.

Therefore, the following sections are intended to introduce the field of magnetism in reduced dimensions, so-called *nanomagnetism*. First, energy contributions are discussed that compete with each other and determine the magnetic state in equilibrium. Highlighted are two distinct magnetic states, namely the single-domain Ising state and the vortex state. The section closes with the discussion of arrays made from such nanomagnetic elements, where the focus lies on uncoupled as well as on highly correlated arrangements.

A detailed step-by-step explanation and an appropriate mathematical description of magnetic order in nanostructures goes beyond the scope of this work. However, a comprehensive presentation of the field of (ferro)magnetism, magnetic materials and nanomagnetism can be found in textbooks [36, 37, 39, 53, 161–165].

2.2.1 Energy Contributions in Micro- and Nanomagnetism

Energy minimisation governs many of the typical length scales that can be found in physical systems and that constrain the systems degrees of freedom. In the case of magnetic materials, the interplay of different local and non-local energy terms defines key properties such as the exchange length, domain-wall width or domain sizes and, further, determines, among others, the magnetic configuration with respect to so-called easy and hard axes and planes. Knowledge and control of these energy terms allows for a precise manipulation of the magnetic state in a nano-, micro-, or macroscopic magnetic specimen and is therefore of great technological importance in, e.g., magnetic memory applications or the optimisation of permanent magnets [53, 162, 166].

Exchange Energy

The exchange interaction is almost always responsible for magnetic order and ferromagnetism in solids. The microscopic coupling mechanism is an interplay of *Coulomb interaction* between electrons on the one hand and the *Pauli principle*, based on the exchange of indistinguishable electrons as fermionic particles, on the other hand [36]. The sign and magnitude of the energy difference between electronic configurations that fulfil a total antisymmetric wave function determines the eventually realised magnetic configuration.
In ferromagnets, a consequence of the exchange interaction is the preferentially collinear alignment of adjacent spins. Any deviation from this uniform alignment results in an energy penalty that is captured as an exchange energy contribution E_{Ex},

$$E_{\text{Ex}} \propto A \int_V (\vec{\nabla}\vec{M})^2 \, dV, \tag{2.9}$$

integrated over the sample volume V, where A is the exchange constant and $\vec{M}$ the magnetisation of the system [53]. The corresponding gain in energy is independent of the direction of the deviating magnetic moments and thus behaves isotropically.
Related to the exchange energy is a characteristic length scale defining the spatial distance over which the parallel alignment due to exchange interaction can be maintained before it gets suppressed or altered by competing interactions, see the following sections. This material-dependent parameter is called the *exchange length* and, for example, is 5.7 nm for bulk permalloy ($Ni_{81}Fe_{19}$) [167], the ferromagnetic material of choice in the present work.

Anisotropy Energy

Magnetic anisotropy describes the energetically favoured (easy axes or easy planes) and unfavoured (hard axes or hard planes) orientations of magnetic moments within a crystal [168]. Often it is separated into a microscopic, material-dependent and a macroscopic, shape-dependent part. While the first part characterises a material's inherent *magnetocrystalline anisotropy*, the second part characterises an anisotropy-

like behaviour originating from shape-determined magnetostatic contributions via the *shape anisotropy* and is, thus, discussed in that particular subsection.
The symmetry of the local environment of magnetic atoms in a crystal determines which electronic orbitals are, due to crystal-field splitting, lower in energy and thus occupied by electrons with—in the sum—-uncompensated magnetic moments. The spin-orbit interaction now leads to preferential orientations of the magnetisation vector with respect to the crystallographic lattice because of the particular spatial alignment of these magnetic-moment carrying orbitals. Note that the concept of magnetocrystalline anisotropy, however, is not restricted to materials with a net magnetisation, but is also applicable to compensated types of magnetic order as well.
In general, the anisotropy energy, E_{A}, can be expressed as a Taylor series containing different orders of the magnetisation $\vec{M}$ integrated over the sample volume V as

$$E_{\text{A}} \propto \int_V \left(K^{(0)} + \sum_i K_i^{(1)} M_i + \sum_{i,j} K_{i,j}^{(2)} M_i M_j + \sum_{i,j,k} K_{i,j,k}^{(3)} M_i M_j M_k + ... \right) dV, \tag{2.10}$$

with $K^{(n)}$ being the anisotropy coefficient or tensors of the order n. For the exemplary case of a crystal with uniaxial anisotropy (i.e. trigonal, tetragonal or hexagonal lattice), the lowest (non-spherical-symmetric) term that contributes to E_{A} is given in spherical coordinates ($M_x = M_0 \cos(\theta)\cos(\phi)$, $M_y = M_0 \cos(\theta)\sin(\phi)$, $M_z = M_0 \sin(\theta)$) by

$$E_{\text{A}} \propto \int_V K^{(2)} M_0^2 \sin^2(\theta)\, dV, \tag{2.11}$$

where $K^{(2)}$ is the second-order anisotropy constant and θ the angle between the magnetisation and the uniaxial axis (z-axis). Depending on the sign of $K^{(2)}$, the preferential alignment of magnetic moments is along the $\pm z$-axis (z: easy axis $\leftrightarrow$ x,y: hard plane) for $K^{(2)} > 0$ or perpendicular to the z-axis (z: hard axis $\leftrightarrow$ x,y: easy plane) for $K^{(2)} < 0$. Higher-order terms with the characteristic of the underlying point-group symmetry may alter the result and induce more complex angular dependences of the anisotropy energy [53].
Of great practical importance is the modification of a ferromagnetic material's global magnetic anisotropy depending on the particular growth method. Here, the magnetocrystalline anisotropy is only distinct in single crystalline materials, whereas for polycrystalline or randomly distributed granular magnetic matter with crystallite sizes of the order of the exchange length the net anisotropy may become negligible, see Ref. [162] p. 855.

Zeeman Energy

As already touched upon in Sect. 2.1.4 on page 16, the direction of magnetisation $\vec{M}$ within a ferromagnetic material reacts upon the application of an external magnetic field. Formally, the external field is introduced as a Zeeman energy, E_{Z}, that scales linear with the magnetic field strength $\vec{H}$ and calculates as an integral over the sample volume V as

$$E_Z \propto -\int_V (\vec{M} \cdot \vec{H})\, dV. \tag{2.12}$$

Consequently, the magnetic moments of the material subjected to the field experience a force that favours an alignment parallel to the external field direction.

Magnetostatic Energy

The total magnetic induction $\vec{B}$ resulting from a magnetic sample with magnetisation $\vec{M}$ in an external field $\vec{H}$ is given by $\vec{B} = \mu_0(\vec{H} + \vec{M})$, with μ_0 being the magnetic permeability of vacuum. Maxwells' equation yields

$$\vec{\nabla}\vec{B} = 0 = \mu_0\vec{\nabla}(\vec{H} + \vec{M}) \quad \longleftrightarrow \quad \vec{\nabla}\vec{M} = -\vec{\nabla}\vec{H}, \tag{2.13}$$

meaning that every divergence in the magnetisation distribution of a sample serves as source or sink for magnetic fields $\vec{H}$. The corresponding field inside the sample is known as the demagnetising field $\vec{H}_d$ and is oriented antiparallel to the internal magnetisation, while the generated field in its surrounding is termed the stray field $\vec{H}_s$. The resulting magnetostatic energy contribution E_{ms} results from the demagnetisation field interacting with the magnetisation of the sample on the one hand and accounts for the generated external magnetic field on the other hand. Separately written, it is given by an integration over the sample volume V_{sample} and by an integration over the sample's exterior $V_{\infty-sample}$ as

$$E_{ms} \propto \int_{V_{\infty-sample}} (\vec{H}_s)^2\, dV - \int_{V_{sample}} (\vec{H}_d\, \vec{M})\, dV \quad . \tag{2.14}$$

In order to minimise magnetostatic energy, the sources and sinks of the demagnetising field have to be minimised. For finite sample volumes this is conterminous with the avoidance of internal magnetisation inhomogeneities (volume magnetic charges) $\sigma_V \propto -\vec{\nabla}\,\vec{M}$, as well as avoidance of magnetisation directions parallel to the sample's surface normal $\vec{n}$, which otherwise yield surface magnetic charges $\sigma_S \propto \vec{n}\,\vec{M}$. This concept is known as the 'pole-avoidance principle' [169]. Practically, to avoid surface magnetic charges, a multi-domain state is formed in which a particular distribution of homogeneously magnetised areas or a continuously rotating magnetic order cancels the total generated magnetic field.
The elaborated factors strongly depend on the shape of a magnetic sample, which significantly influences the possible density of volume- or surface-magnetic charges. This motivates the often-used term *shape anisotropy* for the related energy contribution.

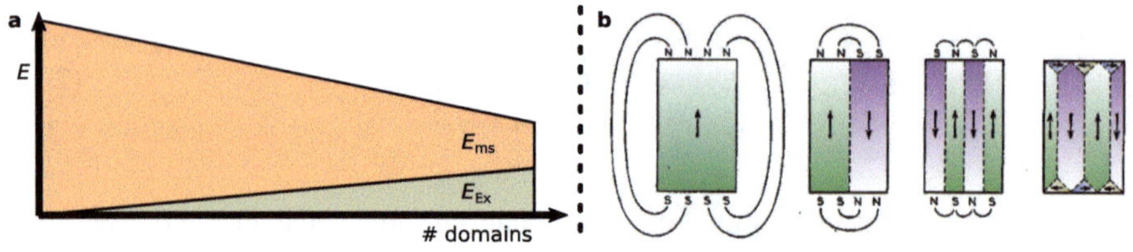

Fig. 2.10 Magnetic configurations in small and thin magnetic elements. a Domain formation in a ferromagnetic specimen causes the consecutive reduction of the internal energy comprising of magnetostatic energy E_{ms} and exchange energy E_{Ex}. **b** Sketch of the transition from a single-domain state with extended magnetic stray fields to a flux-closed multi-domain state. Figure adapted from Ref. [173]

Shape-Dependent Magnetic Configuration

As introduced in the previous sections, the material, shape and dimension of a magnetic object drastically influence the balance of energy terms and, thus, is of major significance for the formation of its internal magnetic configuration [170]. The magnetostatic energy prefers the emergence of magnetic-flux-closed states (multi-domain or non-uniform magnetic states) without internal sinks or sources for magnetic fields. However, associated spin arrangements increase both anisotropy energy and exchange energy due to spins that are neither aligned parallel one to another nor aligned parallel to the system's easy axis or plane.
Assuming negligible magnetocrystalline anisotropy and a uniformly aligned magnetisation $\vec{M}$ over the volume of the sample (in order to neglect volume magnetic charges σ_{V} such that only surface magnetic charges σ_{S} account for the resulting magnetostatic energy), the demagnetising field $\vec{H}_{\mathrm{d}}$ of a magnetic specimen with approximately ellipsoidal shape[6] can be expressed as [172]

$$\vec{H}_{\mathrm{d}} = -N\vec{M} \quad ; \quad N = \begin{pmatrix} N_{xx} & 0 & 0 \\ 0 & N_{yy} & 0 \\ 0 & 0 & N_{zz} \end{pmatrix}, \tag{2.15}$$

where N represents the symmetric *demagnetisation tensor* with a trace of 1. For a thin film ($x, y \to \infty$, $z \to 0$) with z being the surface normal, only N_{zz} exists and equals 1. The resulting shape anisotropy yields an easy plane (xy plane) and, thus, an in-plane magnetisation that avoids large surface magnetic charges. In this case, however, the film may split into a pseudo-random distribution of homogeneously magnetised in-plane domains. The associated gain in exchange energy due to non-parallel-aligned spins at the domain walls can be compensated by the reduction of magnetostatic energy, see Fig. 2.10.
A magnetic multi-domain state is typically realised in bulk materials or patterned films with elements of a lateral size of >2 μm. Evaluating the demagnetisation tensor for more elongated rod-like structures, only two entries (for example

[6] For more realistic shapes such as a ferromagnetic prism, corresponding demagnetisation values have been calculated as well [171].

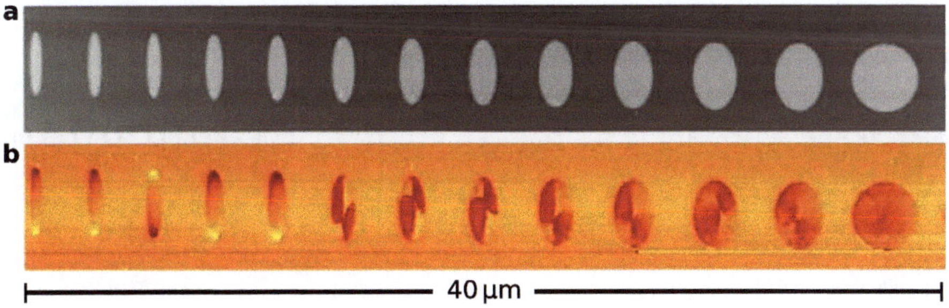

Fig. 2.11 Transition from a single-domain to a magnetic vortex state in nanomagnetic elements with varying aspect ratio. **a** Atomic force microscopy and **b** magnetic force microscopy images of 30-nm-thick polycrystalline NiFe elements. The dark and bright contrast corresponds to surface magnetic charges and, thus, opposite magnetic poles. The length of the vertical axis for each element is 3 μm, while the horizontal elongation is changed to realise aspect ratios from 7.5 (left) to 1 (right). The shape-anisotropy-determined transition from a single-domain state (left) via a four-domain state (centre) to a magnetic vortex state (right) is evident. Figure adapted from Ref. [174]

$N_{yy} \approx N_{zz} \approx 0.5$) exist. The resulting shape anisotropy yields a uniaxial symmetry with an easy axis along x. A ferromagnetic film patterned in such a way behaves like a small rod magnet and exhibits only two stable magnetic states with a magnetisation that is constrained to align parallel to the long axis (here, the x axis). Note that the magnetostatic energy contribution in this case is of minor significance since the total volume of the small specimen results in energetically tolerable surface magnetic charges. These types of nanomagnets, see the left-hand side of Fig. 2.11, can be referred to as *Ising magnets*.

A magnetic element patterned in a more symmetric way may result in a four-domain, so-called *Landau-pattern* or leads to magnetic configurations that do not allow for a description in terms of homogeneous magnetic domains, but as a continuously changing magnetisation distribution forming a magnetic vortex state, see the right-hand side of Fig. 2.11. Here, for a relatively thick disc-like magnetic specimen (and thus weak in-plane shape anisotropy), the delicate interplay between magnetostatic energy (that scales linearly with the film thickness) and exchange energy (as a constant local quantity) yields a magnetic state with a constant non-zero angle between neighbouring magnetic moments. Thus, the energy contribution due to exchange interaction is kept relatively small due to the negligible spin-spin angle on a microscopic scale, whereas the avoidance of magnetisation gradients keeps the magnetostatic energy small. Especially disc-like planar specimens with smooth edges and an aspect ratio close to one allow for the emergence of a magnetic vortex state [175–178], see Fig. 2.12.

Both the single-domain Ising states and the flux-closed vortex states are of particular interest[7] for my work and will be discussed in more detail in the following two subsections.

---

[7] Not discussed here is the rich variety of magnetic configurations that form between these two variants, which have been observed in discs or polyhedral-shaped magnetic bodies, namely the

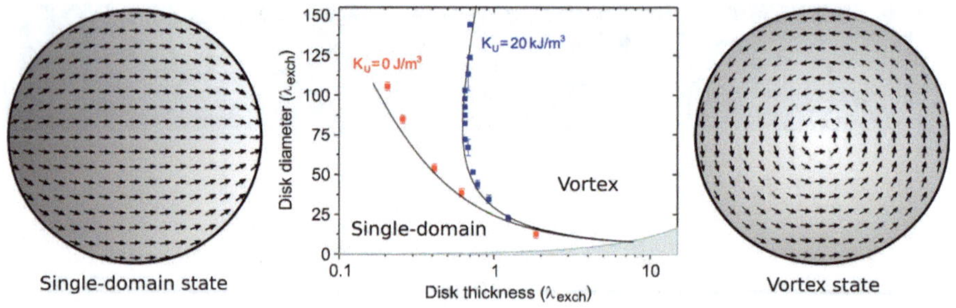

Fig. 2.12 Size-determined transition from the magnetic single-domain to a magnetic vortex state. Depending on the lateral dimensions and the thickness of a disc-shaped nanomagnetic element (in units of the exchange length λ_{exch}) with zero and non-zero uniaxial anisotropy (K_u), a transition from a single-domain state (left) to a vortex state (right) occurs. The two representative spin configurations have been simulated with possibly two different initial magnetisations for a ferromagnetic element with zero in-plane anisotropy, 500 nm diameter and 1 nm thickness. Figure adapted from Ref. [178]

2.2.2 *Ising-Type Nanomagnets: Macrospins*

The macrospin model refers to a magnetic single-domain system, typically with strong uniaxial anisotropy, in which the homogeneous magnetisation distribution can switch coherently between its two stable orientations (its Ising states) along the easy axis: The system reacts analogous to a single giant magnetic moment, therefore termed a macro- or superspin [181]. The remaining degrees of freedom—determined by the anisotropy energy—decide about the particular type of macrospin. For a thin and small disc-shaped magnetic body with zero magnetic anisotropy, see Fig. 2.12, the uniform magnetisation has an in-plane degree of freedom (referred to as the XY model discussed in two-dimensional magnetism), whereas a shape-induced or magnetocrystalline uniaxial in-plane anisotropy reduces the effective degrees of freedom and yields a two-state Ising-like system. For the second case, the magnetisation-reversal mechanism for different angles between the easy axis and the applied field was first described in 1948 as a coherent reorientation of the spin system and can be described using the Stoner-Wohlfarth model [167, 182]. Two approaches for the experimental realisation of an Ising-like nanomagnetic system can be distinguished: a magnetic specimen with a strong uniaxial magnetocrystalline anisotropy or a prolate magnetic particle, both below the critical size for providing a uniform single-domain state.

For a spherical or elliptical disc below a critical size, the minimisation of internal energy is governed by a minimised exchange energy favouring a domain-wall-free single-domain state [183–186]. Two size limits exist: For the upper critical size, the magnetostatic energy gains significance and enforces a multi-domain equilibrium

C state (buckle state), the S state, the triangle state, the flower state, the leaf state or the diamond state [162, 179, 180].

state. For the lower critical size, the thermal energy induces random fluctuations of the magnetisation distribution, referred to as the *superparamagnetic* regime [187]. The name reflects the fact that a corresponding magnetic particle behaves paramagnetically without a remanent net magnetisation, whereas, upon application of an external magnetic field, a huge paramagnetic susceptibility shows up due to the field-stabilised single-domain state originating from the Zeeman-energy contribution, see Sect. 2.2.1. For a particle in the superparamagnetic state, the energy barrier that separates the two Ising states is below the thermal energy. Therefore, thermally induced reorientation events follow the behaviour of activation-energy-related physical effects and scale exponentially with the volume-dependent energy barrier. The average time between magnetisation-reorientation events is called the Néel relaxation time τ_N that follows the famous Néel-Arrhenius law [188].

$$\tau_N \propto \exp\left(\frac{E_A V}{k_B T}\right), \tag{2.16}$$

where $E_A V$ is the energy-barrier height, with E_A being the anisotropy-energy density (see Sect. 2.2.1) and V the volume, and $k_B T$ is the thermal energy, with k_B being the Boltzmann constant and T the temperature. Related to the relaxation time and actual measurement time is the so-called the *blocking temperature* [189] at which the direction of magnetisation of a (superpara-)magnetic body gets virtually frozen. The relaxation time equals the measurement time at this temperature and indicates the experimentally sensed transition from the superparamagnetic regime to the ferromagnetic single-domain regime. This phase transition is passed through during the growth of confined magnetic structures, see Sect. 3.4.

The fabrication of small ferromagnetic structures (see Sect. 3.1 on page 58) provides the flexibility for shaping the energy landscape of magnetic particles and for defining either room-temperature superparamagnetic or blocked/frozen systems on demand. Macrospins can be fabricated by patterning small rod-like magnetic objects from a soft-magnetic material [190, 191], see the right-hand side of Fig. 2.11. In the so-called *macrospin approximation*, a simplistic model for ensembles of magnetic-dipole-interacting macrospins, the net magnetic moment of an extended single-domain magnet can formally be condensed into a point at its geometric centre defining the position of a giant classical 'spin'. The magnetic moment of such a macrospin μ_{mac} of volume V made from soft-ferromagnetic permalloy ($Ni_{81}Fe_{19}$) can be calculated as

$$\mu_{mac} = \mu_{py}\, n_{py}\, V = \mu_{py} \frac{\bar{\rho}_{py}}{\bar{A}_{py}} V, \tag{2.17}$$

where $\mu_{py} \approx 1\,\mu_B$ is the magnetic moment of an average atom in the alloy [192], $n_{py} \approx 90\,\text{nm}^{-3}$ is the effective atomic density and therefore the ratio of averaged mass density $\bar{\rho}_{py} \approx 8.7\,\text{g/cm}^3$ and averaged atomic mass $\bar{A}_{py} \approx 58.1\,\text{u}$. Typical values for μ_{mac} in nano-patterned magnetic structures are in the order of $10^8\,\mu_B$ [193]. This leads to one of the main reasons for the high research interest in nanomagnetic systems: Magnetic-dipole-interaction energies for spin-spin coupling are usually in the range of 1 meV (corresponding to temperatures of $\approx 10\,\mathrm{K}$) and are, thus, in almost all cases not the main source of magnetic order [181]. However, for classical 'spins' exhibiting a six to nine times larger magnetic moment, magnetostatic interactions may become dominant and therefore allow for driving the formation of types of magnetic order in correlated 'spin' systems as it will be seen in the after next section.

2.2.3 Vortex-Type Nanomagnets

For nanomagnetic objects of sizes above the exchange-energy-dominated single-domain limit, inhomogeneous magnetic states form to reduce the magnetostatic energy with flux-closed configurations that reflect the topology and shape of the magnetic specimen. Here, the delicate competition between the different energy contributions of comparable strength offers a strongly textured energy landscape for the spin system to locate its global energy minimum. Besides the possibility of forming a pseudo-random multi-domain state, a continuously varying in-plane magnetisation in thin films of low intrinsic anisotropy may balance and minimise the energy terms by the formation of a magnetic vortex state as a most symmetric variant typically occurring in sub-micrometre-sized magnetic discs [175, 194]. Such a curled spin structure emerges in confined and relatively symmetric geometries, is a very stable topologically non-trivial object [195] and exhibits no in-plane net magnetisation. From topological considerations, however, it follows that a singularity forms in the centre of this vortex structure at which the spins tilt out-of-plane, giving rise to the existence of a spatially highly confined magnetised vortex core [196]. The size of this vortex core lies in the range of the exchange length (see Sect. 2.2.1) and is balanced by the reduction in exchange energy and the gain in magnetostatic energy due to its existence. Since the vortex core is accompanied by an energy contribution due to the surface magnetic charges associated with it, it reduces the overall stability of the vortex state. This energy contribution can be avoided by fabricating the particular magnetic object with a hole at the position where the vortex core is expected [197]. Accordingly, the magnetisation distribution of such a ring-like planar specimen is to a first approximation free of sinks and sources of magnetic fields and, thus, exhibit negligible magnetostatic energy.

The magnetic vortex state has been investigated in planar ferromagnetic objects of different shape, see Ref. [162] p. 851, among them, in circular discs [175, 194, 198, 199], rings [197, 199–201] and polygons [199, 202–209]. As discussed in the next section, the choice of a particular geometry suitable for hosting a remanent magnetic vortex state is of key importance in the context of independently addressable and isolated elements on the one hand, or stray-field-coupled and correlated ensembles of elements in densely packed vortex arrays on the other hand. A magnetic specimen that exhibits a magnetic vortex state inherently provides a giant toroidal moment, see Sect. 2.1.5 on page 17, where the direction of the toroidal moment vector is given by the sense of circulation of the internal magnetic configuration. Thus, coupled magnetic-vortex states appear to be ideal candidates for the study of ferrotoroidicity on mesoscopic length scales.

2.2.4 Arrays of Nanomagnetic Elements

Ensembles of nanomagnetic elements are often arranged in a way to either study effects related to their interaction with each other or to investigate their suitability to be used in densely packed memory arrays. Both of these scenarios are accompanied by different implications about their desired collective behaviour as discussed in the following.

Weakly Interacting Magnetic Arrays

For non-volatile memory applications such as in magnetic hard-disc drives, it is crucial to ensure the thermal stability and to avoid crosstalk between individual magnetic bits that encode binary information. Data is typically encoded in the two Ising states of a magnetic bit on a homogeneous film with a large magnetocrystalline anisotropy. Furthermore, efforts have been made to encode data on patterned magnetic structures [164, 210–215]. The reading and writing process is conveniently realised with a movable head that combines a magnetoresistive sensor for the read-out with a small magnetic field coil for manipulating the magnetisation. A major drawback of Ising-state-based magnetic data storage are the magnetostatic conditions within and in between individual bits. The magnetostatic energy defines a lower-size limit due to the superparamagnetic effect that is approached towards a maximised memory density. Furthermore, the magnetisation of each bit it is the source for an unwanted magnetostatic coupling and hence a crosstalk between neighbouring bits. While the first challenge can be addressed by increasing the anisotropy of the memory layer to stabilise the magnetic state, see Eq. 2.16, the second challenge requires resorting towards stray-field-reduced magnetic elements.
Here, *VRAM* (vortex magnetic random access memory) has been discussed as a high-density magnetic data storage [216] that allows the encoding of binary information in the handedness of the magnetic vortex state [217, 218]. Moreover, a four-state multi-bit memory has been realised using the handedness (clockwise and counter-clockwise whirled magnetic state) on the one hand and the magnetisation direction of the vortex core (up- or down-magnetised state) on the other hand [219, 220]. Because of the inherently weak coupling between the elements, crosstalk is negligible by design. Control over and read-out scenarios for magnetic vortex states have been proposed and experimentally realised by resonantly exciting a vortex-core gyration mode via radio frequency pulses [207, 219–228], via spin-transfer torque [220, 229–231], by a piezoelectric approach in a magnetostriction-coupled magnetoelectric heterostructure [232], or by a geometrical [197, 200, 203, 205, 206, 209, 233, 234], field-induced [199] or interaction-mediator-based [235] asymmetry for accessing the vortex state with homogeneous fields. The applicability of VRAM, however, is not yet on an industrial level of production. Though, the encoding of binary data in (macroscopic) magnetic vortex states is a relatively old concept and has been the basis of so-called *ferrite core memory* invented and used since the 1950s to 1970s.

Strongly Interacting Magnetic Arrays

Densely packed magnetic elements interact with each other, form collective energy states and require a description beyond that of individual magnetic building blocks. In such arrays, also referred to as *artificial spin systems*, first introduced in 2006 [193, 236], a particular microscopic interaction on the atomic scale, e.g. magnetic exchange interaction, is substituted and mimicked by a corresponding magnetic-dipole interaction [237–242] on mesoscopic length scales. In this regard, for the study of a macroscopic phenomenon such as ferromagnetism, the underlying microscopic mechanism is irrelevant as long as the macroscopic characteristic is indistinguishable from the 'atomic prototype'.

The unparalleled advantage of artificial spin systems is the ability to design, engineer and tune inter-'atomic' couplings on demand by fabrication techniques that have matured over the past few decades, see Sect. 3.1. In addition, experimental access to all relevant spin degrees of freedom is possible using a broad variety of methods, see Sect. 3.2, whereas just an average or integrated response is typically probed in atomic systems. The study of macrospin-based arrays has proven useful to address fundamental questions in magnetism [243–246]. Here, the extended magnetic-dipole-like stray fields of macrospins can connect individual building blocks with their neighbours. The pairwise interaction-energy contribution E_D from point-like adjacent magnetic moments is given by

$$E_\mathrm{D} = \frac{\mu_0}{4\pi}\left(\frac{\vec{m}_i\,\vec{m}_j}{|\vec{r}_i - \vec{r}_j|^3} - \frac{3\,(\vec{m}_i\,(\vec{r}_i - \vec{r}_j))\,(\vec{m}_j\,(\vec{r}_i - \vec{r}_j))}{|\vec{r}_i - \vec{r}_j|^5}\right), \qquad (2.18)$$

where μ_0 is the magnetic permeability of vacuum and $\vec{m}_{i,j}$ and $\vec{r}_{i,j}$ are magnetic-moment vectors and spatial positions, respectively, of the ith and jth macrospin. Accordingly, the sign and magnitude of the dipole-interaction energy sensitively depend on the distance and relative orientation of neighbouring macrospins. For two adjacent macrospins with magnetic moments that synchronously rotate in their shared plane, Eq. 2.18 can be evaluated in angular coordinates to quantify the spatial dependence of their interaction,

$$E_\mathrm{D}(\theta) = \frac{\mu_0}{4\pi}\frac{|\vec{m}_i|\,|\vec{m}_j|}{|\vec{r}|^3}\,(1 - 3\cos^2(\theta)) \equiv E_\mathrm{DC}\,(1 - 3\cos^2(\theta)), \qquad (2.19)$$

with θ being the angle of synchronous rotation and E_DC the magnetic-dipole coupling constant. Figure 2.13a shows the transition from a parallel (ferromagnetic) alignment to an antiparallel (antiferromagnetic) alignment upon synchronous rotation of both macrospins. At the so-called *magic angle* of $\theta = \arccos(\sqrt{1/3}) \approx 55°$, derived from Eq. 2.19, the interaction energy vanishes and the two possible magnetic states become degenerate. Furthermore, three selections for two oppositely rotating magnets are shown in Fig. 2.13b.

For more than two macrospins, as for example in the case of arrangements in lattices with fixed positions and two possible orientations of individual magnetic moments, it

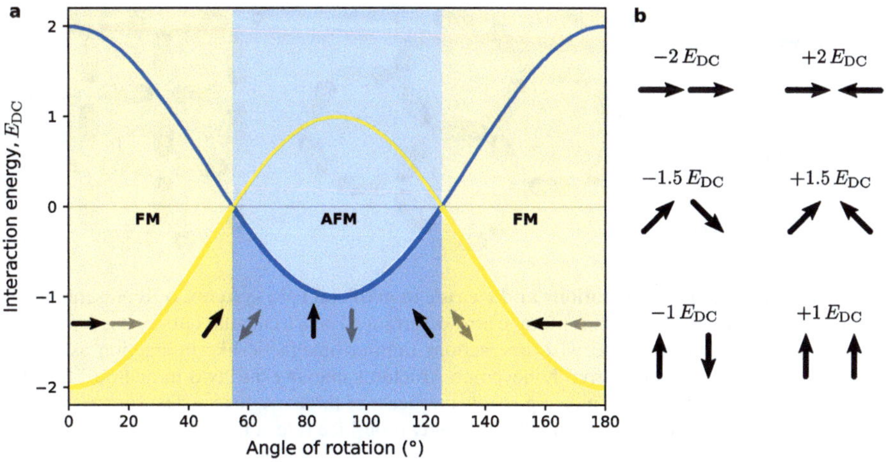

Fig. 2.13 Anisotropy of the magnetic-dipole interaction. The sign and magnitude of the pairwise coupling of magnetic moments, expressed in terms of E_{DC} (see text), sensitively depends on their separation and relative orientation. **a** Interaction energy and transition between compensated (AFM, blue shading) and non-compensated (FM, yellow shading) alignments depending on the angle of rotation. Five representative orientations are sketched as pairs of arrows (black: fixed magnetisation, grey: preferred alignment relative to the neighbour). **b** Interaction energy for three selected configurations for two oppositely rotating magnetic moments (left: head-to-tail, right: head-to-head). The energetically favoured alignments (left-hand side) mimic ferromagnetism (top), antiferromagnetism (bottom) and weak ferromagnetism, also called 'canted antiferromagnetism' (centre)

can be impossible to simultaneously satisfy all configurational constraints at neighbouring lattice sites at which macrospins pairwise strive for minimising their interaction energy. Thus, not a single spin configuration will form a unique global ground state, but several configurations share approximately the same energy. This scenario of a large low-energy-state degeneracy and zero-temperature entropy is termed *frustration* [247–249] and is depicted in Fig. 2.14a.

Considering many-body nanomagnetic systems, the majority of studies has been made on two frustrated spin-system geometries: the kagome- and the square-ice array [164, 243, 244, 246, 250], see Fig. 2.14b, c. Such systems were first introduced to mimic either the geometric frustration in the binary hydrogen-bonding-length degree of freedom in water ice and in the spin frustration in rare-earth titanate pyrochlores $RE_2Ti_2O_7$ [251–254]. Since then, the kagome and the square ice have turned out to be valuable testing grounds to model and study additional very fundamental physical phenomena. Among these are magnetic correlations, frustration and disorder [193, 255–262], the emergence of magnetic monopoles [263–272], the thermal relaxation behaviour [273–275], phase transitions and thermal fluctuations [269, 275–287], approaches for chirality control [288] and the identification of long-range order [275, 289–291]. Externally applied magnetic fields as well as annealing protocols have been employed to modify the magnetic configuration of these patterns [266, 275, 278, 287, 292–295].

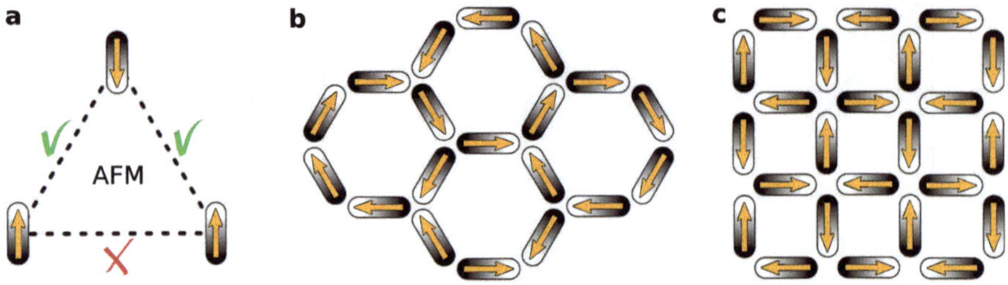

Fig. 2.14 Frustrated configurations and ice rule in artificial spin systems. a Representation of geometrically frustrated Ising spins (orange arrows) arranged on a triangular lattice. It is impossible to simultaneously satisfy all pairwise interactions (antiferromagnetic-like interaction assumed). **b** The kagome-ice array built from honeycomb structures obeying the 'two-in/one-out' or 'one-in/two-out' ice rule [296] at the meeting points (vertices) of three magnets. **c** The square-ice array consisting of square unit cells obeying the 'two-in/two-out' ice rule at the vertices

In order to separate the terms *long-range order* and *primary ferroic order*, a detailed look at the corresponding domain patterns is needed. The domain states in spin-ice systems, see Fig. 2.15a, are *translational domains*, see Ref. [5], p. 40. This term refers to their microscopic condition that a mere translation by a structural unit-cell matches the magnetic pattern of one domain state with the other. If magnetic domain states locally differ by a translation of one structural unit cell or, equivalently, half a magnetic unit cell (which means that the magnetic unit cell is twice the size of the structural unit cell), the considered system are macroscopically invariant under time reversal symmetry [42]. Accordingly, the magnetic configurations forming the two-fold degenerate long-range-ordered states can not be associated to a physical property that uniquely relates to the particular magnetic configuration in either of the two domain states, since both domain types are macroscopically indistinguishable. As a consequence, domain states in kagome-ice or square-ice systems are not accompanied by a primary ferroic order parameter. However, on the microscopic level, the order is made up from magnetic whirls, defining local toroidal moments, that alternate in their sense of circulation such that the net toroidisation (Eq. 2.7) vanishes within one magnetic unit cell, see Fig. 2.15a. Therefore, the spin-arrangement in artificial square ice may be related with an *antiferrotoroidic* state, see Ref. [43] p. 11. This is in analogy with an antiferromagnetic state in which—although magnetically ordered—the net magnetisation cancels at the level of a unit cell.
To the best of my knowledge, this consideration about the non-primary-ferroic nature of spin-ice systems holds true as well for all demonstrated systems with geometric modifications of the conventional ice-type structures, as introduced above. In this context, a variety of geometries have been studied that exhibits more, less or different couplings between macrospins. Among these are asymmetric Ising-like systems with an as-grown ordering and emergent magnetic monopoles [297–300], quasi-crystalline arrays with local transient states and frustration [301–303], ordering in glass-like arrays [304], the 'tetris', 'shakti', 'cairo', 'brickwork' or rewritable-square-ice lattice that exhibit reduced dimensionality, phase transitions, topological

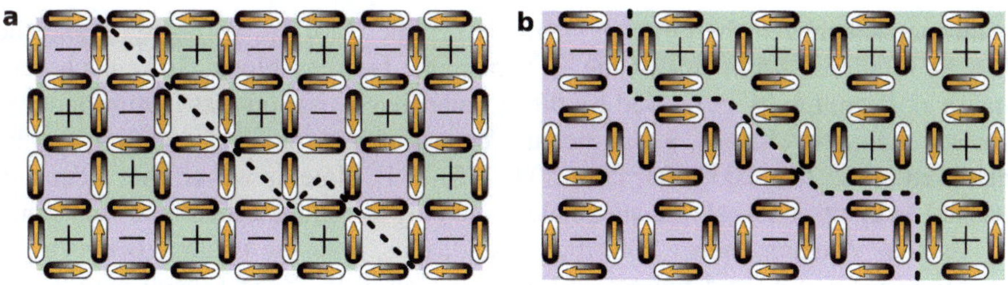

Fig. 2.15 Antiferrotoroidic and ferrotoroidic order in artificial nanomagnetic arrays. a Long-range order in artificial square ice can be classified as antiferrotoroidic order. It is formed by alternating magnetic vortices (toroidal moments, violet: $-t$, green: $+t$) at the unit-cell scale and, thus, yields zero net toroidisation. **b** Long-range ferroic order in the toroidal square array exhibits orientational domains that uniquely relate to the orientation of toroidal moments. These toroidal moments are coupled to collectively form areas with a non-zero toroidisation

measures and local manipulability [250, 259, 293, 305–310], arrays with an emergent chirality and artificial ferromagnetism [311–313], implementations of artificial ferrimagnetism [314], the presence of a phase coexistence described in relation to the Potts model [315, 316], 'anti'-spin-ice systems [317, 318], interaction-modified arrays [319, 320], interconnected variants [289, 321–325] or the coupling to additional functional layers [326].

In contrast, the domain states in primary ferroic systems are of orientational nature and therefore called *orientational domains* [5]. Accordingly, these domains can be distinguished by the orientation of a ferroic order parameter, see Sect. 2.1.2 on page 11 that allows for parametrising macroscopic quantities related to the two domain states. One representation of such an artificial ferroic system is shown in Fig. 2.15b. It can be constructed by splitting all spins of the artificial square ice into two parallel neighbours that favour to align antiparallel to each other. The resulting system is referred to as a toroidal square array and it constitutes the main model system that I have studied throughout my doctorate.

Another class of coupled nanomagnetic arrays, which is much less investigated, is based on building blocks that intrinsically exhibit a magnetic vortex state, see Sect. 2.2.3. Here, the density of surface magnetic charges is negligibly small for circular-shaped ferromagnetic discs that contain a magnetic vortex and, hence, weak magnetostatic interaction impede their coupling. Despite the weak coupling of vortex states in circular discs, it has been demonstrated that during magnetisation reversal in a homogeneous external magnetic field, a magnetostatic coupling arises from the induced net magnetisation due to the non-equilibrium position of the vortex core [327–329]. Nevertheless, a modification of the inherent magnetostatic interaction between magnetic building blocks is essential for a coupling of neighbouring vortex states in zero magnetic field. One option for this is the use of a coupling mediator that—placed in between circular-shaped vortex-state elements—perturbs their magnetisation distribution to introduce a handle for an inter-element coupling [235]. Furthermore, by a mere modification of the circular shape to polygons, there is

a mismatch between the circular magnetic symmetry of the vortex state and the broken rotational symmetry of the ferromagnetic building block. Since each geometrical discontinuity perceived by the circular magnetisation distribution induces surface magnetic charges, a significant increase of inter-element coupling can be achieved [197, 200, 203, 205, 206, 209, 233, 234]. Note, however, that a densely packed array of nanoscale magnetic elements may introduce a local and very inhomogeneous Zeeman energy contribution and, thus, alter the balance between the energy terms that were discussed in Sect. 2.2.1 on page 28. The result may be the destabilisation of the vortex state in favour of a magnetic single-domain state [330, 331]. This is one of the reasons, why studies on collective magnetic order in arrays made from vortex-state magnetic elements are so rare.
Within this work, I used arrays comprising of triangular-shaped ferromagnetic building blocks. This particular shape enables the emergence of a magnetic vortex state, while retaining relatively strong magnetic stray fields to connect neighbouring building blocks.

2.3 Metamaterials and Photonic Crystals

The propagation of light into and through homogeneous media of constant wave impedance has been first studied centuries ago by Christiaan Huygens, Isaac Newton and Augustin-Jean Fresnel and can be described in terms of wave- or simplified ray-optical approaches. If a material is, however, structured with a period of the order of the light wavelength or much below that, advanced concepts may be required for an appropriate description of light-matter interaction. Whereas optical resonances in conventional bulk crystals are determined by electronic energy levels and selection rules corresponding to their particular quantum state, the optical responses of micro- and nanostructured surfaces are typically determined by processes of a different origin. In this regard, novel phenomena of light propagation can be observed and tailored [332] such as negative refraction [333–337], ultra-slow light propagation [338] or a strong localisation and enhancement of electromagnetic fields [339–342].
The microscopic constituents of a patterned material can be separated into two categories: dielectric (insulating) [343–345] and metallic (conducting) [340, 346–348] structures. While dielectric building blocks benefit from a low-loss propagation of radiation and, thus, low extinction coefficients, structures made from electrically conducting building blocks allow for the resonant excitation of collective oscillations of free charge carriers, called *localised surface plasmon polaritons* or *particle plasmon polaritons* [340, 349–351] that are confined to the surface of the metallic particle at which the real part of the dielectric function changes its sign. Another type of resonance occurs due to the periodicity of the structure and is termed *lattice resonance* or *Rayleigh anomaly* [352–357]. It occurs when a wave diffracts into the structure plane and changes its character from radiative to evanescent and—in turn—can couple back to the evanescent field of periodically located particle-resonances. The lattice resonance, primarily depending on the lattice constant, leads to multiple

coherent scattering. Excited by the electric field of the incident radiation at optical, infrared or microwave frequencies, these resonances are accompanied by a confinement of the light field as well as an anomalous dispersion in the dielectric function. As a consequence, light-matter interaction gets enhanced and the sensitivity for probing, for example, an otherwise undetectably small response that originates from an emergent symmetry breaking of the sample increases. Combining this plasmonic character with magnetic materials paves the way for the research field of so-called *magneto-plasmonics* with the main goal to control magnetic order or to enhance the optical response of magnetic materials [358–361].
Even though the nomenclature in literature is not consistent for the description of nano- and microscopic structures, two classes are generally considered:

- Photonic crystals [362–364] are periodically structured materials made from basis units of similar size as compared with the wavelength of the radiation that interacts with it, which is often infrared or visible light. In analogy to Bloch waves of electrons that describe the band structure of electronic states in ordinary crystals, a band structure of photonic states can result from a periodic modulation of the dielectric function. The name therefore relates to physical responses upon incident radiation that is phenomenologically equivalent to the interaction of x-rays or electrons with 'atomic' crystals. The behaviour of photonic crystals is typically dominated by interference effects and diffraction instead of by intrinsic properties of the structure's constituents. Well-known natural examples of photonic crystals can be found in the iridescent-colourful surfaces [365, 366] of some species of insects, for instance in the deep colours of the wings of some butterflies such as individuals from the Lycaenidae family, as well as for the variety of colours found in natural opal or nacre. In a similar way, but with a very different outcome, nanostructures in biologic tissue such as the eye of flies or moths act as antireflection coatings and therefore enhance the visual sensitivity of the insect.
- Metamaterials [220, 367–376] are assembled materials that have a structural period much smaller than that of the radiation that interacts with it, which is for practical reasons often in the microwave-frequency range. As a main consequence, the light field is not subject to interference phenomena but senses the composite material in the *dipole approximation* as a homogeneous effective medium defined by the arrangement of the 'meta' unit cell [339, 377–380]. Alongside the mere description of structures on a meta level, the term *meta* implies material responses that may overcome limitations of ordinary crystalline materials, as briefly overviewed above.

Keywords and terms from literature such as metasurfaces, metacrystals, artificial superlattices or nanoantenna arrays, can be classified accordingly into the two above-mentioned categories. For convenience, the term 'metamaterial' will be mainly used in the following for all kinds of artificial nanostructured materials regardless of their precise classification.
The artificial periodic arrays of magnetic nano-elements that I used for this work, see Sect. 4.1 on page 81, constitute a variant of two-dimensional magneto-photonic crystals in which the optical response of the system may be governed by the lattice

symmetry and the unit-cell assembly. Further, diffraction as such has a big impact on the inherent extinction properties due to the large contrast in refractive indices and, thus, wave impedance between metallic (gold- or aluminium-coated permalloy) nanomagnets and semiconducting silicon as substrate material.

Besides in my main experimental structures, I touched photonic crystals and metamaterials in two further aspects. First, in the context of teaching, I developed, tested and published an experimental setup based on an active photonic crystal, namely a programmable spatial light modulator (SLM), to interactively teach crystal-diffraction [381], see also Sect. A.1 on page 151. Furthermore, the presented setup has been used beyond its teaching capabilities for providing insight into potential scattering phenomena that arise from the interaction of light with my nanomagnetic arrays. The results have inspired the image post-processing algorithm as introduced in Sect. 3.6 on page 75 and the study of optical effects based on toroidal order, see Chap. 7 on page 133.

Second, my peripheral project comprises measurements on natural and artificial biologic nanocoatings. I worked on several insect-eye structures using optical- (reflection spectroscopy) and opto-mechanical (adhesion force microscopy) analysis techniques in collaboration with Mikhail Kryuchkov from the DEPARTMENT OF CELL PHYSIOLOGY AND METABOLISM of the University of Geneva. This work [382–384], however, lies outside of the scope of my main topic and I refrain from a more comprehensive statement at this point.

References

1. Aizu K (1970) Possible species of ferromagnetic, ferroelectric, and ferroelastic crystals. Phys Rev B 2(3):754–772. https://doi.org/10.1103/PhysRevB.2.754
2. Newnham R, Cross L (1974) Symmetry of secondary ferroics. I. Mater Res Bullet 9(7):927-933. https://linkinghub.elsevier.com/retrieve/pii/002554087490172X
3. Newnham R, Cross L (1974) Symmetry of secondary ferroics. II. Mater Res Bullet 9(8):1021–1031. https://doi.org/10.1016/0025-5408(74)90012-9
4. Wadhawan VK (2000) Introduction to ferroic materials. 1st edn. CRC Press. https://doi.org/10.1201/9781482283051
5. Tagantsev AK, Cross LE, Fousek J (2010) Domains in ferroic crystals and thin films. Springer, New York. https://doi.org/10.1007/978-1-4419-1417-0
6. Curie P (1894) Sur la symetrie dans les phenomenes physiques, symetrie d'un champ electrique et d'un champ magnetique. Jde Phys Theor et Appl 3(1):393–415. https://doi.org/10.1051/jphystap:018940030039300
7. Birss RR (1966) Symmetry and magnetism. North-Holland
8. Argyei AK (1992) Phase-transition-related crystal properties: the Neumann principle revisited. J Phys: Condensed Matter 4(36):7499–7514. https://doi.org/10.1088/0953-8984/4/36/021
9. http://reference.iucr.org/dictionary/Neumann's_principle . [Online content, accessed on 22.12.2019]
10. Birss RR (1963) Macroscopic symmetry in space-time. Rep Prog Phys 26(1):307–360. https://doi.org/10.1088/0034-4885/26/1/309
11. http://www.cryst.ehu.es/cryst/mpoint.html . [Online content, accessed on 29.12.2019]
12. http://mpg.web.cmu.edu/ . [Online content, accessed on 22.12.2019]

13. Perez-Mato J et al (2015) Symmetry-based computational tools for magnetic crystallography. Ann Rev Mater Res 45(1):217–248
14. Hlinka J (2014) Eight types of symmetrically distinct vectorlike physical quantities. Phys Rev Lett 113(16):165502. https://doi.org/10.1103/PhysRevLett.113.165502
15. Grimmer H (1994) The forms of tensors describing magnetic, electric and toroidal properties. Ferroelectr 161(1):181-189. https://www.tandfonline.com/doi/abs/10.1080/00150199408213366
16. Schmid H (2001) On ferrotoroidics and electrotoroidic, magnetotoroidic and piezotoroidic effects. Ferroelectrics 252(1):41–50. https://www.tandfonline.com/doi/abs/10.1080/00150190108016239
17. Schmid H (2008) Some symmetry aspects of ferroics and single phase multiferroics. J Phys: Condens Matter 20(43):434201. https://doi.org/10.1088/0953-8984/20/43/434201
18. Litvin DB (2008) Ferroic classifications extended to ferrotoroidic crystals. Acta Crystallogr Sect A Found Crystallogr 64(2):316–320. https://doi.org/10.1107/S0108767307068262
19. Kleber W et al (2010) Einfuehrung in die Kristallographie, vol 19, verb. De Gruyter, Oldenbourg
20. Borchardt-Ott W, Sowa H (2018) Kristallographie. Springer, Berlin
21. Dzyaloshinskii IE (1960) On the magneto-electrical effects in antiferromagnets. J Exp Theoret Phys 10(3):628–629
22. Landau LD, Lifshitz EM, Pitaevskij LP (1984) Electrodynamics of continuous media, vol 8, 2nd edn. Butterworth, Oxford
23. Agyei AK, Birman JL (1990) On the linear magnetoelectric effect. J Phys: Cond Matter 2(13):3007–3020. https://doi.org/10.1088/0953-8984/2/13/010
24. Nye JF (1984) Physical properties of crystals: their representation by tensors and matrices. Clarendon Press, Oxford University Press
25. Newnham RE (2005) Properties of materials: anisotropy, symmetry, structure. Oxford University Press
26. Litvin DB (2010) Ferroic crystals and tensor distinction. Phase Trans 83(9):682–693. https://doi.org/10.1080/01411594.2010.500217
27. https://it.iucr.org/Db/ . [Online content, accessed on 22.12.2019]
28. Aizu K (1973) Second-order ferroic state shifts. J Phys Soc Jap 34(1):121–128
29. Van Aken BB et al (2007) Observation of ferrotoroidic domains. Nature 449(7163):702–705. https://doi.org/10.1038/nature06139
30. Wadhawan VK (1982) Ferroelasticity and related properties of crystals. Phase Trans 3(1):3-103. https://www.tandfonline.com/doi/abs/10.1080/01411598208241323
31. Salje EK (2012) Ferroelastic materials. Ann Rev Mater Res 42(1):265–283. https://doi.org/10.1146/annurev-matsci-070511-155022
32. Cheong S-W et al (2018) Broken symmetries, non-reciprocity, and multiferroicity. Npj Quantum Mater 3(1):19. https://doi.org/10.1038/s41535-018-0092-5
33. Jin W et al (2019) Observation of a ferro-rotational order coupled with second-order nonlinear optical fields. Nat Phys. https://doi.org/10.1038/s41567-019-0695-1
34. Fiebig M (2019) Order! Order!!. Nat Phys (2019). https://doi.org/10.1038/s41567-019-0714-2
35. Rabe KM, Ahn CH, Triscone J-M (2007) Physics of ferroelectrics: a modern perspective, vol 105. Springer, Berlin. https://doi.org/10.1007/978-3-540-34591-6
36. Blundell S (2001) Magnetism in condensed matter. Oxford University Press
37. Coey JMD (2010) Magnetism and magnetic materials. 1st edn. Cambridge University Press. http://dx.doi.org/10.1017/CBO9780511845000
38. Spaldin NA (2011) Magnetic materials: fundamentals and applications. 2nd edn. Cambridge University Press
39. Lueken H et al (1999) Magnetochemie. Vieweg+Teubner Verlag. https://doi.org/10.1007/978-3-322-80118-0

40. Artamonov YA, Gorbatsevich AA, Kopaev YV (1984) Magnetoferroelectric and toroidal ordering. J Exp Theoret Phys 40(7):290–293
41. Dubovik V, Tugushev V (1990) Toroid moments in electrodynamics and solid-state physics. Phys Rep 187(4):145–202. https://doi.org/10.1016/0370-1573(90)90042-Z
42. Ederer C, Spaldin NA (2007) Towards a microscopic theory of toroidal moments in bulk periodic crystals. Phys Rev B 76(21):214404. https://doi.org/10.1103/PhysRevB.76.214404
43. Spaldin NA, Fiebig M, Mostovoy M (2008) The toroidal moment in condensed-matter physics and its relation to the magnetoelectric effect. J Phys: Condens Matter 20(43):434203. https://doi.org/10.1088/0953-8984/20/43/434203
44. Delaney KT, Mostovoy M, Spaldin NA (2009) Superexchange-driven magnetoelectricity in magnetic vortices. Phys Rev Lett 102(15):157203. https://doi.org/10.1103/PhysRevLett.102.157203
45. Spaldin NA et al (2013) Monopole-based formalism for the diagonal magnetoelectric response. Phys Rev B 88(9):094429. https://doi.org/10.1103/PhysRevB.88.094429
46. Izyumov YA, Syromyatnikov VN (1990) Phase transitions and crystal symmetry. Springer, Netherlands
47. Landau LD (1937) On the theory of phase transitions. J Exp Theoret Phys 7:19–32
48. Lenz W (1920) Beitrag zum Verstaendnis der magnetischen Erscheinungen in festen Koerpern. Physikalische Zeitschrift 21:613–615. https://doi.org/10.1038/srep32864
49. Ising E (1925) Beitrag zur theorie des ferromagnetismus. Zeitschrift fuer Physik 31(1):253–258. https://doi.org/10.1007/BF02980577
50. Niss M (2005) History of the Lenz-Ising Model 1920–1950: from ferromagnetic to cooperative phenomena. Arch Hist Exact Sci 59(3):267–318. https://doi.org/10.1007/s00407-004-0088-3
51. Gallavotti G (1999) Statistical mechanics. Springer, Berlin. https://doi.org/10.1007/978-3-662-03952-6
52. Onsager L (1944) Crystal Statistics. I. A two-dimensional model with an order-disorder transition. Phys Rev 65(3–4):117–149. https://doi.org/10.1103/PhysRev.65.117
53. Hubert A, Schaefer R (2009) Magnetic domains: the analysis of magnetic microstructures. Springer
54. Seul M, Andelman D (1995) Domain shapes and patterns: the phenomenology of modulated phases. Science 267(5197):476–483. https://doi.org/10.1126/science.267.5197.476
55. Farztdinov MM (1965) Structure of antiferromagnets. Sov Phys Uspekhi 7(6):855-876. https://doi.org/10.1070/PU1965v007n06ABEH003689
56. Fiebig M et al (1998) Second harmonic spectroscopy and control of domain size in antiferromagnetic $YMnO_3$. J Appl Phys 83(11):6560–6562. https://doi.org/10.1063/1.367579
57. Li Y-Y (1956) Domain walls in antiferromagnets and the weak ferromagnetism of α-Fe_2O_3. Phys Rev 101(5):1450–1454. https://doi.org/10.1103/PhysRev.101.1450
58. Catalan G et al (2012) Domain wall nanoelectronics. Rev Mod Phys 84(1):119–156. https://doi.org/10.1103/RevModPhys.84.119
59. Meier D (2015) Functional domain walls in multiferroics. J Phys: Condens Matter 27(46):463003. https://doi.org/10.1088/0953-8984/27/46/463003
60. Selzer S et al (2016) Inertia-free thermally driven domain-wall Motion in antiferromagnets. Phys Revi Lett 117(10):107201. https://doi.org/10.1103/PhysRevLett.117.107201
61. Seidel J (2016) Topological structures in ferroic materials. Springer, Berlin
62. Chen S et al (2020) Recent progress on topological structures in ferroic thin films and heterostructures. Adv Mater 2000857. https://doi.org/10.1002/adma.202000857
63. Dzyaloshinskii IE (1957) Thermodynamical theory of weak ferromagnetism in antiferromagnetic substances. J Exper Theoret Phys 5(6):1259–1272
64. Moriya T (1960) Anisotropic superexchange interaction and weak ferromagnetism. Phys Rev 120(1):91–98. https://doi.org/10.1103/PhysRev.120.91
65. Bode M et al (2007) Chiral magnetic order at surfaces driven by inversion asymmetry. Nature 447(7141):190–193. https://doi.org/10.1038/nature05802

66. Chen G et al (2013) Novel chiral magnetic domain wall structure in Fe/Ni/Cu(001) films. Phys Rev Lett 110(17):177204. https://www.nature.com/articles/ncomms15302
67. Emori S et al (2013) Current-driven dynamics of chiral ferromagnetic domain walls. Nat Mater 12(7):611–616. https://doi.org/10.1038/nmat3675
68. Benitez MJ et al (2015) Magnetic microscopy and topological stability of homochiral Néel domain walls in a Pt/Co/AlOx trilayer. Nat Commun 6(1):8957. https://doi.org/10.1038/ncomms9957
69. Soumyanarayanan A et al (2016) Emergent phenomena induced by spin-orbit coupling at surfaces and interfaces. Nature 539(7630):509–517. https://www.nature.com/articles/nature19820
70. Chen G et al (2017) Out-of-plane chiral domain wall spin-structures in ultrathin in-plane magnets. Nature Commun 8(1):15302. https://www.nature.com/articles/ncomms15302
71. Artamonov YA, Gorbatsevich AA (1985) Symmetry and dynamics of systems with toroidal moments. J Exper Theoret Phys 89:1078–1092
72. Papasimakis N et al (2016) Electromagnetic toroidal excitations in matter and free space. Nat Mater 15(3):263–271. https://doi.org/10.1038/nmat4563
73. Nanz S (2016) Toroidal multipole moments in classical electrodynamics: an analysis of their emergence and physical significance. Springer, Berlin
74. Talebi N, Guo S, van Aken PA (2018) Theory and applications of toroidal moments in electrodynamics: their emergence, characteristics, and technological relevance. Nanophotonics 7(1):93–110. https://doi.org/10.1515/nanoph-2017-0017
75. Gorbatsevich AA, Kopaev YV (1994) Toroidal order in crystals. Ferroelectrics 161(1):321-334. https://www.tandfonline.com/doi/abs/10.1080/00150199408213381
76. Kopaev YV (2009) Toroidal ordering in crystals. Physics-Uspekhi 52(11):1111–1125. https://doi.org/10.3367/UFNe.0179.200911d.1175
77. Planes A, Castan T, Saxena A (2015) Recent progress in the thermodynamics of ferrotoroidic materials. Multiferroic Mater 1(1):9–22. https://doi.org/10.2478/muma-2014-0002
78. Gnewuch S, Rodriguez EE (2019) The fourth ferroic order: Current status on ferrotoroidic materials. J Solid State Chem 271:175–190. https://doi.org/10.1016/j.jssc.2018.12.035
79. Zel'dovich IB (1957) Electromagnetic interaction with parity violation. J Exper Theoret Phys 1184–1186
80. Ginzburg V et al(1984) On the problem of superdiamagnetism. Solid State Commun 50(4):339-343. https://doi.org/10.1016/0038-1098(84)90381-8
81. Tokura Y (2000) Orbital physics in transition-metal oxides. Science 288(5465):462-468. https://doi.org/10.1126/science.288.5465.462
82. Scagnoli V et al (2011) Observation of Orbital Currents in CuO. Science 332(6030):696–698. https://doi.org/10.1126/science.1201061
83. Naumov II, Bellaiche L, Fu H (2004) Unusual phase transitions in ferroelectric nanodisks and nanorods 432:4
84. Prosandeev S et al (2006) Controlling toroidal moment by means of an inhomogeneous static field: an Ab initio study. Phys Rev Lett 96(23):237601. https://doi.org/10.1103/PhysRevLett.96.237601
85. Guo LY et al (2012) Electric toroidal dipole response in split-ring resonator metamaterials. Eur Phys J B 85(6):208. https://doi.org/10.1140/epjb/e2012-20935-3
86. Dubovik VM, Tosunyan LA, Tugushev VV (1986) Axial toroidal moments in electrodynamics and solid-state physics, p 8
87. Batista CD, Ortiz G, Aligia AA (2008) Ferrotoroidic moment as a quantum geometric phase. Phys Rev Lette 101(7):077203. https://doi.org/10.1103/PhysRevLett.101.077203
88. Sugano S, Kojima N (2000) Magneto-optics. Springer
89. Froehlich D et al (1999) Nonlinear spectroscopy of antiferromagnetics. Appl Phys B: Lasers Opt 68(3):465–471. https://doi.org/10.1007/s003400050650
90. Pisarev RV, Fiebig M, Froehlich D (1997) Nonlinear optical spectroscopy of magnetoelectric and piezomagnetic crystals. Ferroelectrics 204(1):1–21. https://www.tandfonline.com/doi/abs/10.1080/00150199708222184

91. Sherstyuk NE et al (2015) Optical second harmonic generation microscopy for ferroic materials. Ferroelectrics 477(1):29–46. https://www.tandfonline.com/doi/full/10.1080/00150193.2015.999597
92. Fiebig M, Pavlov VV, Pisarev RV (2005) Second-harmonic generation as a tool for studying electronic and magnetic structures of crystals: review. J Opt Soc Ame B 22(1):96
93. Tasset F (1989) Zero field neutron polarimetry. Phys B: Cond Matter 627–630. https://doi.org/10.1016/S0921-4526(99)00029-0
94. Tasset F et al (1999) Spherical neutron polarimetry with Cryopad-II. Phys B: Condense Matter 267–268:69–74. https://doi.org/10.1016/j.solidstatesciences.2004.11.014
95. Brown PJ, Forsyth JB, Tasset F (2005) Studies of magneto-electric crystals using spherical neutron polarimetry. Solid State Sci 7(6):682–689
96. Janoschek M et al (2007) Spherical neutron polarimetry with MuPAD. Phys B: Condense Matter 397(1):125–130
97. Brown WF, Shtrikman S, Treves D (1963) Possibility of visual observation of antiferromagnetic domains. J Appl Phys 34(4):1233-1234. https://doi.org/10.1016/j.solidstatesciences.2004.11.014
98. Folven E, Takamura Y, Grepstad JK (2012) X-PEEM study of antiferromagnetic domain patterns in $LaFeO_3$ thin films and embedded nanostructures. J Electron Spectrosc Related Phenomena 185(10):381–388. https://doi.org/10.1016/j.elspec.2012.07.001
99. Toledano P et al (2015) Primary ferrotoroidicity in antiferromagnets. Phys Rev B 92(9):094431. https://doi.org/10.1103/PhysRevB.92.094431
100. Zimmermann AS, Meier D, Fiebig M (2014) Ferroic nature of magnetic toroidal order. Nat Commun 5(1):4796. https://doi.org/10.1038/ncomms5796
101. Ascher E (1974) Kineto-electric and Kinetomagnetic effects in crystals. Int J Magnet 5:287–295
102. Freeman AJ, Schmid H, Institute BM (eds) (1975) Magnetoelectric interaction phenomena in crystals. Gordon and Breach Science Publishers
103. Rivera J-P (1994) The linear magnetoelectric effect in $LiCoPO_4$ revisited. Ferroelectrics 161(1):147-164. https://www.tandfonline.com/doi/abs/10.1080/00150199408213364
104. Kornev I et al (2000) Magnetoelectric properties of $LiCoPO_4$ and $LiNiPO_4$. Phys Rev B 62(18):12247–12253. https://doi.org/10.1103/PhysRevB.62.12247
105. Vaknin D et al (2002) Weakly coupled antiferromagnetic planes in single-crystal $LiCoPO_4$. Phys Rev B 65(22):224414. http://doi.org/10.1103/PhysRevB.65.224414
106. Kharchenko YN et al (2003) Weak ferromagnetism and an intermediate incommensurate antiferromagnetic phase in $LiNiPO_4$. Low Temp Phys 29(7):579–583
107. Toft-Petersen R et al (2015) Anomalous magnetic structure and spin dynamics in magnetoelectric LiFePO4. Phys Rev B 92(2):024404
108. Remeika JP (1960) $GaFeO_3$: a ferromagnetic-piezoelectric compound. J Appl Phys 31(5):263–264
109. Rado GT (1964) Observation and possible mechanisms of magnetoelectric effects in a ferromagnet. Phys Rev Lett 13(10):335–337
110. Popov YF et al (1998) Magnetoelectric effect and toroidal ordering in $Ga_{2-x}Fe_xO_3$. J Exper Theoret Phys 87(1):146–151. https://doi.org/10.1134/1.558635
111. Kubota M et al (2004) X-Ray directional dichroism of a polar ferrimagnet. Phys Rev Lett 92(13):137401. https://doi.org/10.1103/PhysRevLett.92.137401
112. Jung JH et al (2004) Optical magnetoelectric effect in the polar $GaFeO_3$ ferrimagnet. Phys Rev Lett 93(3):037403. https://doi.org/10.1103/PhysRevLett.93.037403
113. Arima T et al (2004) Structural and magnetoelectric properties of $Ga_{2-x}Fe_xO_3$ single crystals grown by a floating-zone method. Phys Rev B 70(6):064426. http://dx.doi.org/10.1103/PhysRevB.70.064426
114. Ferromagnetic-type order of atomic multipoles in the polar ferrimagnetic $GaFeO_3$. Phys Rev B 85(14):144421. https://doi.org/10.1103/PhysRevB.85.144421
115. Lee C et al (2014) Analysis of the difference between the pyroxenes $LiFeSi_2O_6$ and $LiFeGe_2O_6$ in Their Spin Order, Spin orientation, and ferrotoroidal order. Chem Mater 26(4):1745–1750. https://doi.org/10.1021/cm5001413

116. Baum M et al (2013) Controlling toroidal moments by crossed electric and magnetic fields. Phys Rev B 88(2):024414. https://doi.org/10.1103/PhysRevB.88.024414
117. Popov YF et al (2001) Features of the magnetoelectric properties of $BiFeO_3$ in high magnetic fields. Low Temp Phys 27(6):478-479. http://aip.scitation.org/doi/10.1063/1.1382990
118. Fiebig M, Eremenko VV, Chupis IE (eds) (2004) Magnetoelectric interaction phenomena in crystals. Kluwer Academic Publishers
119. Ederer C (2009) Toroidal moments as indicator for magneto-electric coupling: the case of $BiFeO_3$ versus $FeTiO_3$. Eur Phys J B 71(3):349–354. https://doi.org/10.1140/epjb/e2009-00274-4
120. Sannikov DG (1997) Phenomenological theory of the magnetoelectric effect in some boracites. J Exper Theoret Phys 84(2):293–299
121. Sannikov DG (1998) Ferrotoroic phase transition in boracites. Ferroelectrics 219(1):177–181. https://doi.org/10.1134/1.558116
122. Ressouche E et al (2010) Magnetoelectric $MnPS_3$ as a candidate for ferrotoroidicity. Phys Rev B 82(10):100408. https://doi.org/10.1103/PhysRevB.82.100408
123. Popov YF et al (1999) Magnetic-field-induced toroidal moment in the magnetoelectric Cr_2O_3. J Exper Theoret Phys Lett 69(4):330–335. https://doi.org/10.1134/1.568032
124. Luzon J et al (2008) Spin Chirality in a molecular dysprosium triangle: the archetype of the noncollinear ising model. Phys Rev Lett 100(24):247205. https://doi.org/10.1103/PhysRevLett.100.247205
125. Soncini A, Chibotaru LF (2008) Toroidal magnetic states in molecular wheels: interplay between isotropic exchange interactions and local magnetic anisotropy. Phys Rev B 77(22):220406. https://doi.org/10.1103/PhysRevB.77.220406
126. Popov AI, Plokhov DI, Zvezdin AK (2009) Anapole moment and spin-electric interactions in rare-earth nanoclusters. Europhys Lett 87(6):67004. https://doi.org/10.1209/0295-5075/87/67004
127. Hewitt IJ et al (2010) Coupling Dy3 triangles enhances their slow magnetic relaxation. Angew Chem Int Edn 49(36):6352–6356. https://doi.org/10.1002/anie.201002691
128. Ungur L et al (2012) Net toroidal magnetic moment in the ground state of a Dy6- triethanolamine ring. J Am Chem Soc 134(45):18554–18557
129. Ungur L et al (2014) Single-molecule toroics in Ising-type lanthanide molecular clusters. Chem Soc Rev 43(20):6894–6905. https://doi.org/10.1021/ja309211d
130. Guo P-H et al (2012) The first Dy4 single-molecule magnet with a toroidal magnetic moment in the ground state. Inorgan Chem 51(3):1233–1235
131. Plokhov DI, Zvezdin AK, Popov AI (2011) Macroscopic quantum dynamics of toroidal moment in Ising-type rare-earth clusters. Phys Rev B 83(18):184415
132. Plokhov DI, Popov AI, Zvezdin AK (2011) Quantum magnetoelectric effect in the molecular crystal Dy3. Phys Rev B 84(22):224436
133. Rivera J-P (2009) A short review of the magnetoelectric effect and related experimental techniques on single phase (multi-) ferroics. Eur Phys J B 71(3):299-313. https://doi.org/10.1140/epjb/e2009-00336-7
134. Pyatakov AP, Zvezdin AK (2012) Magnetoelectric and multiferroic media. Physics- Uspekhi 55(6):557–581. https://doi.org/10.3367/UFNe.0182.201206b.0593
135. Dong S et al (2015) Multiferroic materials and magnetoelectric physics: symmetry, entanglement, excitation, and topology. Adv Phys 64(5):519–626. https://doi.org/10.1080/00018732.2015.1114338
136. Fiebig M et al (2016) The evolution of multiferroics. Nat Rev Mater 1(8):16046. https://doi.org/10.1038/natrevmats.2016.46
137. Fiebig M (2005) Revival of the magnetoelectric effect. J Phys D: Appl Phys 38(8):123–152
138. Gorbatsevich AA, Omel'yanovskii OE, Tsebro VI (2009) Toroidal ordering in crystals and nanostructures. Physics-Uspekhi 52(8):835–845
139. Zelezny J et al (2014) Relativistic Néel-order fields induced by electrical current in antiferromagnets. Phys Rev Lett 113(15):157201. https://doi.org/10.1103/PhysRevLett.113.157201

140. Gomonay EV, Loktev VM (2014) Spintronics of antiferromagnetic systems (Review Article). Low Temperat Phys 40(1):17–35. https://doi.org/10.1063/1.4862467
141. Jungwirth T et al (2016) Antiferromagnetic spintronics. Nat Nanotechnol 11(3):231–241. https://doi.org/10.1038/s41567-018-0063-6
142. Wadley P et al (2016) Electrical switching of an antiferromagnet. Science 351(6273):587–590. https://doi.org/10.1126/science.aab1031
143. Godinho et al (2018) Electrically induced and detected Néel vector reversal in a collinear antiferromagnet. Nat Commun 9(1):4686. https://doi.org/10.1126/science.aab1031
144. Baltz V et al (2018) Antiferromagnetic spintronics. Rev Mod Phys 90(1):015005. https://doi.org/10.1103/RevModPhys.90.015005
145. Astrov DN (1960) The magnetoelectric effect in antiferromagnetics. J Exper Theoret Phys 11(3):708–709
146. Rado GT, Folen VJ (1962) Magnetoelectric effects in antiferromagnetics. In: Proceedings of the seventh conference on magnetism and magnetic materials. In: Osborn JA (ed) Springer, pp 1126–1132. https://link.springer.com/article/10.1007
147. Suryanarayana SV (1994) Magnetoelectric interaction phenomena in materials. Bull Mater Sci 17(7):1259-1270. https://link.springer.com/article/10.1007
148. Hu J-M, Nan C-W (2019) Opportunities and challenges for magnetoelectric devices. APL Mater 7(8):080905. https://doi.org/10.1063/1.5112089
149. Eerenstein W, Mathur ND, Scott JF (2006) Multiferroic and magnetoelectric materials. Nature 442(7104):759-765. https://www.nature.com/articles/nature05023
150. Lawes G, Srinivasan G (2011) Introduction to magnetoelectric coupling and multiferroic films. J Phys D: Appl Phys 44(24):243001. https://iopscience.iop.org/article/10.1088/0022-3727/44/24/243001
151. Tokura Y, Seki S, Nagaosa N (2014) Multiferroics of spin origin. Rep Prog Phys 77(7):076501. https://www.nature.com/articles/s41467-018-05759-4
152. Wang X et al (2015) Observation of magnetoelectric multiferroicity in a cubic perovskite system: $LaMn_3Cr_4O_{12}$. Phys Rev Lett 115(8):087601. https://doi.org/10.1103/PhysRevLett.115.087601
153. Zhou L et al (2017) Realization of large electric polarization and strong magnetoelectric coupling in $BiMn_3Cr_4O_{12}$. Adv Mater 29(44):1703435. https://doi.org/10.1002/adma.201703435
154. Shen X et al (2019) Large linear magnetoelectric effect and field-induced ferromagnetism and ferroelectricity in $DyCrO_4$. NPG Asia Mater 11(1):50. https://www.nature.com/articles/s41427-019-0151-9
155. Zhou G et al (2019) Large magnetoelectric effect in the polar magnet Sm_2BaCuO_5. Appl Phys Lett 115(25):252902. https://doi.org/10.1063/1.5127893
156. Sawada K, Nagaosa N (2005) Optical magnetoelectric effect in multiferroic materials: evidence for a lorentz force acting on a ray of light. Phys Rev Lett 95(23):237402. https://doi.org/10.1103/PhysRevLett.95.237402
157. Arima T (2008) Magneto-electric optics in non-centrosymmetric ferromagnets. J Phys: Condens Matter 20(43):434211. https://doi.org/10.1088/0953-8984/20/43/434211
158. Kida N et al (2005) Optical magnetoelectric effect in a submicron patterned magnet. Phys Rev Lett 94(7):077205. https://doi.org/10.1103/PhysRevLett.94.077205
159. Caloz C et al (2018) Electromagnetic nonreciprocity. Phys Rev Appl 10(4):047001. https://doi.org/10.1103/PhysRevApplied.10.047001
160. Watanabe H, Yanase Y (2018) Symmetry analysis of current-induced switching of antiferromagnets. Phys Rev B 98(22):220412. https://doi.org/10.1103/PhysRevB.98.220412
161. Stoehr J, Siegmann HC (2006) Magnetism: from fundamentals to nanoscale dynamics, vol 152. Springer
162. Kronmueller H, Parkin S (eds) Handbook of magnetism and advanced magnetic materials. 1st edn. Wiley. https://onlinelibrary.wiley.com/doi/book/10.1002/9780470022184
163. Aktas B, Tagirov L, Mikailov F (eds) Magnetic nanostructures, vol 94. Springer (2007)

164. Zabel H, Farle M (eds) Magnetic nanostructures: spin dynamics and spin transport. vol 246. Springer (2013)
165. Fermon C, Voorde MHVD (eds) (2017) Nanomagnetism: applications and perspectives. Wiley-VCH Verlag
166. Stamps RL et al (2014) The 2014 Magnetism roadmap. J Phys D: Appl Phys 47(33):333001
167. Tannous C, Gieraltowski J (2008) The stoner-wohlfarth model of ferromagnetism. Eur J Phys 29(3):475–487. https://doi.org/10.1088/0143-0807/29/3/008
168. Yosida K (1968) The status of the theories of magnetic anisotropy. J Appl Phys 39(2):511–518. https://doi.org/10.1063/1.2163506
169. Aharoni A (1996) Introduction to the theory of ferromagnetism. Oxford University Press
170. Chien CL, Zhu FQ, Zhu J-G (2007) Patterned nanomagnets. Phys Today 60(6):40–45. physicstoday.scitation. https://doi.org/10.1063/1.2754602
171. Aharoni A (1998) Demagnetizing factors for rectangular ferromagnetic prisms. J Appl Phys 83(6):3432–3434. https://doi.org/10.1063/1.367113
172. Osborn JA (1945) Demagnetizing factors of the general ellipsoid. Phys Rev 67(11):351–357. https://doi.org/10.1103/PhysRev.67.351
173. Kittel C (1949) Physical theory of ferromagnetic domains. Rev Mod Phys 21(4):541–583. https://doi.org/10.1103/RevModPhys.21.541
174. Imre A et al (2006) Majority logic gate for magnetic quantum-dot cellular automata. Science 311(5758):205–208. https://doi.org/10.1126/science.1120506
175. Cowburn RP et al (1999) Single-domain circular nanomagnets. Phys Rev Lett 83(5):1042–1045. https://doi.org/10.1103/PhysRevLett.83.1042
176. Hoffmann H, Steinbauer F (2002) Single domain and vortex state in ferromagnetic circular nanodots. J Appl Phys 92(9):5463–5467. https://doi.org/10.1063/1.1510559
177. Last T et al (2004) Optimization of nanopatterned permalloy electrodes for a lateral hybrid spin-valve structure. J Appl Phys 96(11):6706–6711. https://doi.org/10.1063/1.1810197
178. Jubert P-O, Allenspach R (2004) Analytical approach to the single-domain-to-vortex transition in small magnetic disks. Phys Rev B 70(14):144402. https://doi.org/10.1103/PhysRevB.70.144402
179. Cowburn RP, Welland M (1998) Phase transitions in planar magnetic nanostructures. Appl Phys Lett 72(16):2041–2043. https://doi.org/10.1063/1.121258
180. Wurft T et al (2019) Evolution of magnetic vortex formation in micron-sized disks. Appl Phys Lett 115(13). https://doi.org/10.1063/1.5116299
181. Bedanta S, Kleemann W (2009) Supermagnetism. J Phys D: Appl Phys 42(1):013001. https://doi.org/10.1088/0022-3727/42/1/013001
182. Stoner EC, Wohlfarth EP (1948) A mechanism of magnetic hysteresis in heterogeneous alloys. 240:44
183. Brown WF (1957) Criterion for uniform micromagnetization. Phys Rev 105(5):1479–1482. https://doi.org/10.1103/PhysRev.105.1479
184. Brown WF (1968) The fundamental theorem of fine-ferromagnetic-particle theory. J Appl Phys 39(2):993–994
185. Aharoni A (1988) Elongated single-domain ferromagnetic particles. J Appl Phys 63(12):5879–5882. https://doi.org/10.1063/1.340280
186. Fratta G, Serpico C, d'Aquino M (2012) A generalization of the fundamental theorem of brown for fine ferromagnetic particles. Phys B: Condensed Matter 407(9):1368–1371. https://doi.org/10.1016/j.physb.2011.10.010
187. Bean CP, Livingston JD (1959) Superparamagnetism. J Appl Phys 30(4):S120–S129. https://doi.org/10.1063/1.2185850
188. Néel L (1950) Theorie du trainage magnetique des substances massives dans le domaine de rayleigh. J Phys et le Radium 11(2):49–61. https://doi.org/10.1051/jphysrad:0195000110204900
189. Bruvera IJ et al (2015) Determination of the blocking temperature of magnetic nanoparticles: the good, the bad, and the ugly. J Appl Phys 118(18):184304. https://doi.org/10.1063/1.4935484

190. Perigo EA et al (2018) Past, present, and future of soft magnetic composites. Appl Phys Rev 5(3):031301. http://aip.scitation.org/doi/10.1063/1.5027045
191. Silveyra JM et al (2018) Soft magnetic materials for a sustainable and electrified world. Science 362(6413):eaao0195. https://doi.org/10.1126/science.aao0195
192. Glaubitz B et al (2011) Development of magnetic moments in $Fe_{1-x}Ni_x$ - alloys. J Phys: Condens Matter 23(25):254210. https://doi.org/10.1088/0953-8984/23/25/254210
193. Wang RF et al (2006) Artificial spinice in a geometrically frustrated lattice of nanoscale ferromagnetic islands. Nature 439(7074):303–306. https://www.nature.com/articles/nature04447
194. Raabe J et al (2000) Magnetization pattern of ferromagnetic nanodisks. J Appl Phys 88(7):4437. https://doi.org/10.1063/1.1289216
195. Thiaville A et al (2003) Micromagnetic study of Bloch-point-mediated vortex core reversal. Phys Rev B 67(9):094410. https://doi.org/10.1103/PhysRevB.67.094410
196. Wachowiak A (2002) Direct Observation of internal spin Structure of magnetic vortex cores. Science 298(5593):577–580. https://doi.org/10.1126/science.1075302
197. Klaeui M et al (2003) Vortex formation in narrow ferromagnetic rings. J Phys: Condens Matter 15(21):985–1024. https://doi.org/10.1088/0953-8984/15/21/201
198. Shinjo T (2000) Magnetic vortex core observation in circular dots of permalloy. Science 289(5481):930–932. https://doi.org/10.1126/science.289.5481.930
199. Konoto M et al (2008) Formation and control of magnetic vortex chirality in patterned micromagnet arrays. J Appl Phys 103(2):023904. https://doi.org/10.1063/1.2828177
200. Klaeui M et al (2001) Vortex circulation control in mesoscopic ring magnets. Appl Phys Lett 78(21):3268–3270
201. Montoncello F et al (2008) Magnetization reversal and soft modes in nanorings: transitions between onion and vortex states studied by Brillouin light scattering. Phys Rev B 78(10):104421. https://doi.org/10.1103/PhysRevB.78.104421
202. Cowburn RP (2000) Property variation with shape in magnetic nanoelements. J Phys D: Appl Phys 33(1):1–16. https://doi.org/10.1088/0022-3727/33/1/201
203. Jaafar M et al (2008) Field induced vortex dynamics in magnetic Ni nanotriangles. Nanotechnology 19(28):285717
204. Lua SYH et al (2009) Spin configuration of hexagonal shaped ferromagnetic elements arranged in different structures. J Appl Phys 105(7):07A319
205. Jaafar M et al (2010) Control of the chirality and polarity of magnetic vortices in triangular nanodots. Phys Rev B 81(5):054439. https://doi.org/10.1103/PhysRevB.81.054439
206. Yakata S et al (2010) Control of vortex chirality in regular polygonal nanomagnets using in-plane magnetic field. Appl Phys Lett 97(22). https://doi.org/10.1063/1.3521407
207. Vogel A et al (2012) Vortex dynamics in triangular-shaped confining potentials. J Appl Phys 112(6):063916
208. Udalov OG et al (2012) Nonreciprocal light diffraction by a lattice of magnetic vortices. Phys Rev B 86(9):094416
209. Krutyanskiy VL et al (2013) Second harmonic generation in magnetic nanoparticles with vortex magnetic state. Phys Rev B 88(9):094424
210. White R, Newt R, Pease R (1997) Patterned media: a viable route to 50 $Gbit/in^2$ and up for magnetic recording? IEEE Trans Magnet 33(1):990–995. https://doi.org/10.1109/20.560144
211. Lohau J et al (2001) Writing and reading perpendicular magnetic recording media patterned by a focused ion beam. Appl Phys Lett 78(7):990–992. https://doi.org/10.1063/1.1347390
212. Richter H, Harkness S (2006) Media for magnetic recording beyond 100 $Gbit/in^2$. MRS Bull 31(5):384–388. https://doi.org/10.1557/mrs2006.98
213. Dobisz E et al (2008) Patterned media: nanofabrication challenges of future disk drives. Proc IEEE 96(11):1836–1846. https://doi.org/10.1109/JPROC.2008.2007600
214. Albrecht TR et al (2015) Bit-patterned magnetic recording: theory, media fabrication, and recording performance. IEEE Trans Magnet 51(5):1–42. https://doi.org/10.1109/TMAG.2015.2397880
215. Tannous C, Comstock RL (2017) Magnetic information-storage materials. In: Kasap S, Capper P (eds) Springer handbook of electronic and photonic materials. Springer International Publishing, p 1

216. Cowburn RP (2007) Change of direction. Nature Mater 6(4):255–256. https://doi.org/10.1038/nmat1877
217. Hertel R (2013) Vortex states a la carte. Nat Nanotechnol 8(5):318-320. https://www.nature.com/articles/nnano.2013.81
218. Zheng Y, Chen WJ (2017) Characteristics and controllability of vortices in ferromagnetics, ferroelectrics, and multiferroics. Rep Prog Phys 80(8). https://doi.org/10.1088/1361-6633/aa5e03
219. Jenkins AS et al (2014) Controlling the chirality and polarity of vortices in magnetic tunnel junctions. Appl Phys Lett 105(17):172403
220. Luo Y et al (2019) Separated edge-soliton-mediated dynamic switching of vortex chirality and polarity. Phys Rev Appl 11(4):044090
221. Guslienko KY et al (2002) Eigenfrequencies of vortex state excitations in magnetic submicronsize disks. J Appl Phys 91(10):8037. https://doi.org/10.1063/1.1450816
222. Van Waeyenberge B et al (2006) Magnetic vortex core reversal by excitation with short bursts of an alternating field. Nature 444(7118):461–464. https://doi.org/10.1038/nature05240
223. Yamada K et al (2007) Electrical switching of the vortex core in a magnetic disk. Nat Mater 6(4):270–273. https://doi.org/10.1038/nmat1867
224. Antos R, Otani Y (2009) Simulations of the dynamic switching of vortex chirality in magnetic nanodisks by a uniform field pulse. Phys Rev B 80(14). https://doi.org/10.1103/PhysRevB.80.140404
225. Jung H et al (2010) Observation of coupled vortex gyrations by 70-ps-time- and 20-nm-spaceresolved full-field magnetic transmission soft x-ray microscopy. Appl Phys Lett 97(22):222502. https://www.nature.com/articles/srep00059
226. Jung H et al (2011) Tunable negligible-loss energy transfer between dipolar-coupled magnetic disks by stimulated vortex gyration. Scient Rep 1(1):59. https://doi.org/https://www.nature.com/articles/srep00059
227. Uhlir V et al (2013) Dynamic switching of the spin circulation in tapered magnetic nanodisks. Nat Nanotechnol 8(5):341–346. https://doi.org/10.1038/nnano.2013.66
228. Yakata S et al (2013) Wide range tuning of resonant frequency for a vortex core in a regular triangle magnet. Sci Rep 3(1):3567. https://doi.org/10.1038/srep03567
229. Kim S-K et al (2008) Reliable low-power control of ultrafast vortex-core switching with the selectivity in an array of vortex states by in-plane circular-rotational magnetic fields and spin-polarized currents. Appl Phys Lett 92(2):022509
230. Sluka V et al (2015) Spin-torque-induced dynamics at fine-split frequencies in nano-oscillators with two stacked vortices. Nat Commun 6(1):6409. https://doi.org/10.1038/ncomms7409
231. Hrkac G et al (2015) Magnetic vortex oscillators. J Phys D: Appl Phys 48(45):453001. https://doi.org/10.1088/0022-3727/48/45/453001
232. Filianina M et al (2019) Piezo-electrical control of gyration dynamics of magnetic vortices. Appl Phys Lett 115(6):062404. https://doi.org/10.1063/1.5110169
233. Schneider M, Hoffmann H, Zweck J (2001) Magnetic switching of single vortex permalloy elements. Appl Phys Lett 79(19):3113–3115. https://doi.org/10.1063/1.1410873
234. Prosandeev S et al (2008) Control of vortices by homogeneous fields in asymmetric ferroelectric and ferromagnetic rings. Phys Rev Lett 100(4):047201. https://doi.org/10.1103/PhysRevLett.100.047201
235. Haldar A, Adeyeye AO (2015) Vortex chirality control in circular disks using dipolecoupled nanomagnets. Appl Phys Lett 106(3):032404
236. Remhof A et al (2007) Dipolar interactions in periodic arrays of rectangular ferromagnetic islands. J Magnet Magnet Mater 310(2):794–796. https://doi.org/10.1016/j.jmmm.2006.11.101
237. Luttinger JM, Tisza L (1946) Theory of dipole interaction in crystals. Phys Rev 70(11):954–964. https://doi.org/10.1103/PhysRev.70.954
238. Cohen MH, Keffer F (1955) Dipolar ferromagnetism at 0K. Phys Rev 99(4):1135–1140. https://doi.org/10.1103/PhysRev.99.1135

239. Theophilou A (1973) The ground state of crystals with dipole-dipole and exchange interactions in an external magnetic field. Physica 70(3):599–615. https://doi.org/10.1016/0031-8914(73)90366-2
240. Niemeijer T, Bloete H (1973) On the ground state of crystals with dipole-dipole and exchange interactions II. Physica 67(1):125–136. https://doi.org/10.1016/0031-8914(73)90027-X
241. Kraemer C et al (2012) Dipolar antiferromagnetism and quantum criticality in $LiErF_4$. Science 336(6087):1416–1419. https://doi.org/10.1126/science.1221878
242. Alkadour B et al (2017) Dipolar ferromagnetism in three-dimensional superlattices of nanoparticles. Phys Rev B 95(21):214407. https://doi.org/10.1103/PhysRevB.95.214407
243. Heyderman LJ, Stamps RL (2013) Artificial ferroic systems: novel functionality from structure, interactions and dynamics. J Phys: Condens Matter 25(36):363201. https://doi.org/10.1088/0953-8984/25/36/363201
244. Nisoli C, Moessner R, Schiffer P (2013) Colloquium: artificial spin ice: designing and imaging magnetic frustration. Rev Mod Phys 85(4):1473–1490. https://doi.org/10.1103/RevModPhys.85.1473
245. Rougemaille N, Canals B (2019) Cooperative magnetic phenomena in artificial spin systems: spin liquids, Coulomb phase and fragmentation of magnetism–a colloquium. Eur Phys J B 92(3):62. https://doi.org/10.1140/epjb/e2018-90346-7
246. Skjaervo SH et al (2020) Advances in artificial spin ice. Nat Rev Phys 2(1):13–28. https://doi.org/10.1088/0022-3727/45/43/433001
247. Toulouse G, Vannimenus J, Maillard J (1977) Spin glasses and roughening transition. J Phys Lett 38(22):459–461. https://doi.org/10.1051/jphyslet:019770038022045900
248. Kirkpatrick S (1977) Frustration and ground-state degeneracy in spin glasses. Phys Rev B 16(10):4630–4641. https://doi.org/10.1103/PhysRevB.16.4630
249. Ramirez AP (1994) Strongly geometrically frustrated magnets. Ann Rev Mater Sci 24(1):453–480. https://doi.org/10.1146/annurev.ms.24.080194.002321
250. Nisoli C (2018) Topology by design in magnetic nano-materials: artificial spin ice. In: Gupta S, Saxena A (eds) The role of topology in materials. vol. 189. Springer International Publishing, pp 85–112
251. Harris MJ et al (1997) Geometrical frustration in the ferromagnetic pyrochlore $Ho_2Ti_2O_7$. Phys Rev Lett 79(13):2554–2557. https://doi.org/10.1103/PhysRevLett.79.2554
252. Ramirez AP et al (1999) Zero-point entropy in 'spin ice'. Nature 399(6734):333–335. https://www.annualreviews.org/doi/10.1146/annurev.ms.24.080194.002321
253. Bramwell ST (2001) Spin ice state in frustrated magnetic pyrochlore materials. Sci 294(5546)(2001):1495-1501. https://doi.org/10.1126/science.1064761
254. Rau JG, Gingras MJ (2019) Frustrated quantum rare-earth pyrochlores. Ann Rev Cond Matter Phys 10(1):357–386. https://doi.org/10.1146/annurev-conmatphys-022317-110520
255. Remhof A et al (2008) Magnetostatic interactions on a square lattice. Phys Rev B 77(13):134409. https://doi.org/10.1103/PhysRevB.77.134409
256. Mengotti E et al (2008) Building blocks of an artificial kagome spin ice: Photoemission electron microscopy of arrays of ferromagnetic islands. Phys Rev B 78(14):144402. https://doi.org/10.1103/PhysRevB.78.144402
257. Li Y, Wang T (2010) A simulation study on the magnetic ordering in an artificial geometrically frustrated lattice. Phys Lett A 374(43):4475–4478. https://doi.org/10.1016/j.physleta.2010.09.010
258. Budrikis Z, Politi P, Stamps RL (2011) Diversity enabling equilibration: disorder and the ground state in artificial spin ice. Phys Rev Lett 107(21):217204. https://doi.org/10.1103/PhysRevLett.107.217204
259. Morrison MJ, Nelson TR, Nisoli C (2013) Unhappy vertices in artificial spin ice: new degeneracies from vertex frustration. New J Phys 15(4):045009. https://doi.org/10.1088/1367-2630/15/4/045009
260. Gliga S et al (2015) Broken vertex symmetry and finite zero-point entropy in the artificial square ice ground state. Phys Rev B 92(6):060413. https://doi.org/10.1103/PhysRevB.92.060413

261. Iacocca E et al (2016) Reconfigurable wave band structure of an artificial square ice. Phys Rev B 93(13):134420. https://doi.org/10.1103/PhysRevB.93.134420
262. Budrikis Z et al (2012) Disorder strength and field-driven ground state domain formation in artificial spin ice: experiment, simulation, and theory. Phys Rev Lett 109(3):037203
263. Castelnovo C, Moessner R, Sondhi SL (2008) Magnetic monopoles in spin ice. Nature 451(7174):42–45. https://doi.org/10.1038/nature06433
264. Mol LA et al (2009) Magnetic monopole and string excitations in two-dimensional spin ice. J Appl Phys 106(6):063913
265. Ladak S et al (2010) Direct observation of magnetic monopole defects in an artificial spin-ice system. Nat Phys 6(5):359–363. https://doi.org/10.1038/nphys1628
266. Mengotti E et al (2011) Real-space observation of emergent magnetic monopoles and associated Dirac strings in artificial kagome spin ice. Nature Phys 7(1):68–74. https://doi.org/10.1038/nphys1794
267. Nascimento FS et al (2012) From confinement to deconfinement of magnetic monopoles in artificial rectangular spin ices. New J Phys 14(11):115019. https://doi.org/10.1088/1367-2630/14/11/115019
268. Rougemaille N et al (2013) Chiral nature of magnetic monopoles in artificial spin ice. New J Phys 15(3):035026
269. Canals B et al (2016) Fragmentation of magnetism in artificial kagome dipolar spin ice. Nat Commun 7(1):11446. https://www.nature.com/articles/ncomms11446
270. Farhan A et al (2017) Magnetic charge and moment dynamics in artificial kagome spin ice. Phys Rev B 96(6):064409. https://doi.org/10.1103/PhysRevB.96.064409
271. Chavez AC, Barra A, Carman GP (2018) Voltage control of magnetic monopoles in artificial spin ice. J Phys D: Appl Phys 51(23):234001. https://doi.org/10.1088/1361-6463/aac0ae
272. Morgan JP et al (2011) Magnetic reversal of an artificial square ice: dipolar correlation and charge ordering. J Phys 13(10):105002
273. Wang RF et al (2007) Demagnetization protocols for frustrated interacting nanomagnet arrays. J Appl Phys 101(9):09J104
274. Ke X et al (2008) Energy minimization and ac demagnetization in a nanomagnet array. Phys Rev Lett 101(3):037205. https://doi.org/10.1103/PhysRevLett.101.037205
275. Farhan A et al (2013) Direct observation of thermal relaxation in artificial spin ice. Phys Rev Lett 111(5):057204
276. Nisoli C (2012) On thermalization of magnetic nano-arrays at fabrication. New J Phys 14(3):035017. https://doi.org/10.1088/1367-2630/14/3/035017
277. Montaigne F et al (2015) Size distribution of magnetic charge domains in thermally activated but out-of-equilibrium artificial spin ice. Scient Rep 4(1):5702.h https://www.nature.com/articles/srep05702
278. Farhan A et al (2014) Thermally induced magnetic relaxation in building blocks of artificial kagome spin ice. Phys Rev B 89(21):214405. https://doi.org/10.1038/nphys2613
279. Anghinolfi L et al (2015) Thermodynamic phase transitions in a frustrated magnetic metamaterial. Nat Commun 6(1):8278. https://www.nature.com/articles/ncomms9278
280. Drisko J, Daunheimer S, Cumings J (2015) $FePd_3$ as a material for studying thermally active artificial spin ice systems. Phys Rev B 91(22):224406. https://doi.org/10.1103/PhysRevB.91.224406
281. Kapaklis V et al (2014) Thermal fluctuations in artificial spin ice. Nat Nanotechnol 9(7):514–519. https://doi.org/10.1038/nnano.2014.104
282. Morley SA et al (2017) Vogel-Fulcher-Tammann freezing of a thermally fluctuating artificial spin ice probed by x-ray photon correlation spectroscopy. Phys Rev B 95(10):104422. https://doi.org/10.1103/PhysRevB.95.104422
283. Sendetskyi O et al (2016) Magnetic diffuse scattering in artificial kagome spin ice. Phys Rev B 93(22):224413. https://doi.org/10.1103/PhysRevB.93.224413
284. Sendetskyi O et al (2019) Continuous magnetic phase transition in artificial square ice. Phys Rev B 99(21):214430. https://doi.org/10.1103/PhysRevB.93.224413

285. Massouras M et al (2020) Probing the antiferromagnetic-paramagnetic transition in artificial spin ice by tuning interactions. Phys Rev B 101(17):174421. https://doi.org/10.1103/PhysRevB.101.174421
286. Budrikis Z et al (2012) Domain dynamics and fluctuations in artificial square ice at finite temperatures. J Phys 14(3):035014
287. Farhan A et al (2013) Exploring hyper-cubic energy landscapes in thermally active finite artificial spin-ice systems. Nat Phys 9(6):375–382
288. Chopdekar RV et al (2013) (2013) Controlling vortex chirality in hexagonal building blocks of artificial spin ice. New J Phy 15(12):125033. https://doi.org/10.1088/1367-2630/15/12/125033
289. Branford WR et al (2012) Emerging chirality in artificial spin ice. Science 335(6076):1597–1600. https://doi.org/10.1126/science.1211379
290. Zhang S et al (2013) Crystallites of magnetic charges in artificial spin ice. Nature 500(7464):553–557. https://doi.org/10.1038/nature12399
291. Morgan JP et al (2011) Thermal ground-state ordering and elementary excitations in artificial magnetic square ice. Nat Phys 7(1):75–79
292. Gartside JC et al (2016) A novel method for the injection and manipulation of magnetic charge states in nanostructures. Sci Rep 6(1):32864. https://www.nature.com/articles/srep32864
293. Wang Y-L et al (2016) Rewritable artificial magnetic charge ice. Science 352(6288):962–966. https://doi.org/10.1126/science.aad8037
294. Gartside JC et al (2018) Realization of ground state in artificial kagome spin ice via topological defect-driven magnetic writing. Nat Nanotech 13(1):53–58. https://doi.org/10.1038/s41565-017-0002-1
295. Parakkat VM, Xie K, Krishnan KM (2019) Tunable ground state in heterostructured artificial spin ice with exchange bias. Phys Rev B 99(5):054429. https://doi.org/10.1103/PhysRevB.99.054429
296. Ortiz-Ambriz A et al (2019) Colloquium: ice rule and emergent frustration in particle ice and beyond. Rev Mod Phys 91(4):041003
297. Loreto RP et al (2015) Emergence and mobility of monopoles in a unidirectional arrangement of magnetic nanoislands. Nanotechnology 26(29):295303. https://doi.org/10.1088/0957-4484/26/29/295303
298. Arnalds UB et al (2016) A new look on the two-dimensional Ising model: thermal artificial spins. New J Phys 18(2):023008. https://doi.org/10.1088/1367-2630/18/2/023008
299. Goncalves RS et al (2019) Tuning magnetic monopole population and mobility in unidirectional array of nanomagnets as a function of lattice parameters. Appl Phys Lett 114(14):142401
300. Loreto RP et al (2019) Experimental and theoretical evidences for the ice regime in planar artificial spin ices. J Phys: Cond Matter 31(2):025301. https://doi.org/10.1088/1361-648X/aaeeef
301. Shi J et al (2018) Plasmonic enhancement and manipulation of optical nonlinearity in monolayer tungsten disulfide. Laser Photon Rev 12(10):1800188. https://doi.org/10.1002/lpor.201800188
302. Brajuskovic V et al (2018) Observation of transient states during magnetization reversal in a quasicrystal artificial spin ice. Phys Rev B 98(9):094424. https://doi.org/10.1103/PhysRevB.98.094424
303. Barrows F et al (2019) Emergent magnetic ordering and topological frustration in quasicrystal artificial spin ices. Phys Rev B 99(9):094424. https://doi.org/10.1103/PhysRevB.99.094424
304. Saccone M et al (2019) Towards artificial Ising spin glasses: Thermal ordering in randomized arrays of Ising-type nanomagnets. Phy Rev B 99(22):224403
305. Gilbert I et al (2016) Emergent reduced dimensionality by vertex frustration in artificial spin ice. Nat Phys 12(2):162–165. https://doi.org/10.1038/nphys3520
306. Gilbert I et al (2014) Emergent ice rule and magnetic charge screening from vertex frustration in artificial spin ice. Nat Phys 10(9):670–675. https://doi.org/10.1038/nphys3037
307. Park J et al (2017) Magnetic response of brickwork artificial spin ice. Phys Rev B 96(2):024436. https://doi.org/10.1103/PhysRevB.96.024436

308. Stopfel H et al (2018) Magnetic order and energy-scale hierarchy in artificial spin-ice structures. Phys Rev B 98(1):014435. https://doi.org/10.1103/PhysRevB.98.014435
309. Stancioli RA, Mol LAS (2019) Intermediate phase and pseudo phase transition in an artificial spin ice model. Phys Rev B 100(2). https://doi.org/10.1103/PhysRevB.100.024432
310. Saccone M et al (2019) Dipolar Cairo lattice: geometrical frustration and short-range correlations. Phys Rev Mater 3(10):104402
311. Gliga S et al (2017) Emergent dynamic chirality in a thermally driven artificial spin ratchet. Nat Mater 16(11):1106–1111. https://doi.org/10.1038/nmat5007
312. Macedo R et al (2018) Apparent ferromagnetism in the pinwheel artificial spin ice. Phys Rev B 98(1). https://doi.org/10.1103/PhysRevB.98.014437
313. Paterson GW et al (2019) Heisenberg pseudo-exchange and emergent anisotropies in field-driven pinwheel artificial spin ice. Phys Rev B 100(17):174410. https://doi.org/10.1103/PhysRevB.100.174410
314. Haldar A, Adeyeye AO (2018) Reconfigurable magnetic and microwave properties of a ferrimagnetic-type artificial crystal. J Appl Phys 123(24). https://doi.org/10.1063/1.5032158
315. Louis D et al (2018) A tunable magnetic metamaterial based on the dipolar four-state Potts model. Nature Mater 17(12):1076–1080. https://doi.org/10.1038/s41563-018-0199-x
316. Sklenar J et al (2019) Field-induced phase coexistence in an artificial spin ice. Nat Phys 15(2):191–195. https://doi.org/10.1038/s41567-018-0348-9
317. Zhou X et al (2016) Large area artificial spin ice and anti-spin ice $Ni_{80}Fe_{20}$ structures: static and dynamic behavior. Adv Funct Mater 26(9):1437-1444. https://onlinelibrary.wiley.com/doi/10.1002/adfm.201505165
318. Mamica S et al (2018) Spin-wave dynamics in artificial anti-spin-ice systems: experimental and theoretical investigations. Phys Rev B 98(5). https://doi.org/10.1103/PhysRevB.98.054405
319. Perrin Y, Canals B, Rougemaille N (2016) Extensive degeneracy, coulomb phase and magnetic monopoles in artificial square ice. Nature 540(7633):410–413. https://doi.org/10.1038/nature20155
320. Oestman E et al (2018) Interaction modifiers in artificial spin ices. Nat Phys 14(4):375–379. https://doi.org/10.1038/s41567-017-0027-2
321. Tanaka M et al (2006) Magnetic interactions in a ferromagnetic honeycomb nanoscale network. Phys Rev B 73(5). https://doi.org/10.1103/PhysRevB.73.052411
322. Burn DM, Chadha M, Branford WR (2017) Dynamic dependence to domain wall propagation through artificial spin ice. Phys Rev B 95(10). https://doi.org/10.1103/PhysRevB.95.104417
323. Bhat VS et al (2016) Magnetization dynamics of topological defects and the spin solid in a kagome artificial spin ice. Phys Rev B 93(14):140401. https://doi.org/10.1103/PhysRevB.93.140401
324. Bhat VS et al (2017) Angular-dependent magnetization dynamics of kagome artificial spin ice incorporating topological defects. Phys Rev B 96(1). https://doi.org/10.1103/PhysRevB.96.014426
325. Perrin Y, Canals B, Rougemaille N (2019) Quasidegenerate ice manifold in a purely two-dimensional square array of nanomagnets. Phys Rev B 99(22). https://doi.org/10.1103/PhysRevB.99.224434
326. Wang Y-L et al (2018) Switchable geometric frustration in an artificial-spin-ice-superconductor heterosystem. Nat Nanotechnol 13(7):560–565. https://doi.org/10.1038/s41565-018-0162-7
327. Zhu X et al (2002) Magnetization reversal and configurational anisotropy of dense permalloy dot arrays. Appl Phys Lett 80(25):4789–4791
328. Natali M et al (2002) Correlated magnetic vortex chains in mesoscopic cobalt dot arrays. Phys Rev Lett 88(15). https://doi.org/10.1103/PhysRevLett.88.157203
329. Natali M et al (2004) Correlated vortex chiralities in interacting permalloy dot patterns. J Appl Phys 96(8):4334–4341
330. Wang J, Adeyeye AO, Singh N (2005) Magnetostatic interactions in mesoscopic $Ni_{80}Fe_{20}$ ring arrays. Appl Phys Lett 87(26):262508

331. Adeyeye AO et al (2007) Magnetostatic coupling in arrays of elongated $Ni_{80}Fe_{20}$ rings. J Phys D: Appl Phys 40(21):6479–6483. https://doi.org/10.1088/0022-3727/40/21/001
332. Veselago VG (1968) The electrodynamics of substances with simultaneously negative values of ϵ and μ. Sov Phys Uspekhi 10(4):509–514. https://doi.org/10.1070/PU1968v010n04ABEH003699
333. Smith DR et al (2000) Composite medium with simultaneously negative permeability and permittivity. Phys Rev Lett 84(18):4184–4187. https://doi.org/10.1126/science.1096796
334. Smith DR (2004) Metamaterials and negative refractive index. Science 305(5685):788–792. https://doi.org/10.1103/PhysRevLett.84.4184
335. Ramakrishna SA (2005) Physics of negative refractive index materials. Rep Prog Phys 68(2):449–521. https://doi.org/10.1088/0034-4885/68/2/R06
336. Pendry J (2004) Negative refraction. Contemporary Phys 45(3):191–202. https://doi.org/10.1080/00107510410001667434
337. Shalaev VM (2007) Optical negative-index metamaterials. Nat Photon 1(1):41-48. https://www.nature.com/articles/nphoton.2006.49
338. Baba T (2008) Slow light in photonic crystals. Nat Photon 2(8):465-473. https://www.nature.com/articles/nphoton.2008.146
339. Pendry J et al (1999) Magnetism from conductors and enhanced nonlinear phenomena. IEEE Trans Microw Theo Tech 47(11):2075–2084. https://doi.org/10.1080/00107510410001667434
340. Maier SA (2007) Plasmonics: fundamentals and applications. Springer
341. Yu H et al (2019) Plasmon-enhanced light-matter interactions and applications. NPJ Comput Mater 5(1):45. https://www.nature.com/articles/s41524-019-0184-1
342. Dombi P et al (2020) Strong-field nano-optics. Rev Mod Phys 92(2):025003. https://doi.org/10.1103/RevModPhys.92.025003
343. Jahani S, Jacob Z (2016) All-dielectric metamaterials. Nat Nanotech 11(1):23–36. https://doi.org/10.1038/nnano.2015.304
344. Baranov DG et al (2017) All-dielectric nanophotonics: the quest for better materials and fabrication techniques. Optica 4(7):814. https://doi.org/10.1364/OPTICA.4.000814
345. Brener I (2020) Dielectric metamaterials fundamentals, designs and applications. Woodhead Publishing
346. Bohm D, Pines D (1951) A collective description of electron interactions. I. Magnetic Interactions. Phys Rev 82(5):625–634. https://doi.org/10.1103/PhysRev.82.625
347. Pines D, Bohm D (1952) A collective description of electron interactions: II. Collective vs individual particle aspects of the interactions. Phys Rev 85(2):338-353. https://doi.org/10.1103/PhysRev.85.338
348. Bohm D, Pines D (1953) A collective description of electron interactions: III. Coulomb interactions in a degenerate electron gas. Phys Rev 92(3):609-625. https://journals.aps.org/pr/abstract/10.1103/PhysRev.92.609
349. Rycenga M et al (2011) Controlling the synthesis and assembly of silver nanostructures for plasmonic applications. Chem Rev 111(6):3669–3712. https://doi.org/10.1021/cr100275d
350. Zhang J, Zhang L, Xu W (2012) Surface plasmon polaritons: physics and applications. J Phys D: Appl Phys 45(11):113001. https://doi.org/10.1088/0022-3727/45/11/113001
351. Maradudin AA, Sambles R, Barnes WL (2014) Modern plasmonics. Elsevier
352. Carron KT et al (1986) Resonances of two-dimensional particle gratings in surface-enhanced Raman scattering. J Opt Soc Ame B 3(3):430. https://doi.org/10.1364/JOSAB.3.000430
353. Auguie B et al (2010) Diffractive arrays of gold nanoparticles near an interface: critical role of the substrate. Phys Rev B 82(15):155447. https://doi.org/10.1103/PhysRevB.82.155447
354. Liu W (2016) Optical resonance problem in metamaterial arrays: a lattice dynamics approach. J Phys: Condensed Matter 28(47):475401. https://doi.org/10.1088/0953-8984/28/47/475401
355. Kravets VG et al (2018) Plasmonic surface lattice resonances: a review of properties and applications. Chem Rev 118(12):5912-5951. https://doi.org/10.1021/acs.chemrev.8b00243
356. Castellanos GW, Bai P, Gomez Rivas J (2019) Lattice resonances in dielectric metasurfaces. J Appl Phys 125(21):213105

357. Goerlitzer ESA et al (2020) Chiral surface lattice resonances. Adv Mater 32(22):2001330. https://onlinelibrary.wiley.com/doi/10.1002/adma.202001330
358. Armelles G et al (2013) Magnetoplasmonics: combining magnetic and plasmonic functionalities. Adv Opt Mater 1(1):10–35. https://doi.org/10.1002/adom.201200011
359. Maksymov IS (2016) Magneto-plasmonic nanoantennas: basics and applications. Rev Phys 1:36–51. https://doi.org/10.1016/j.revip.2016.03.002
360. Pineider F, Sangregorio C (2018) Nanomaterials for magnetoplasmonics. In: Novel magnetic nanostructures. Elsevier, pp 191–220
361. Maccaferri N et al (2020) Nanoscale magnetophotonics. J Appl Phys 127(8):080903
362. Yablonovitch E, Gmitter TJ (1989) Photonic band structure: the face-centered-cubic case. Phys Rev Lett 63(18):1950–1953. https://doi.org/10.1103/PhysRevLett.63.1950
363. Ho KM, Chan CT, Soukoulis CM (1990) Existence of a photonic gap in periodic dielectric structures. Phys Rev Lett 65(25):3152–3155. https://doi.org/10.1103/PhysRevLett.65.3152
364. Sievenpiper DF, Sickmiller ME, Yablonovitch E (1996) 3D wire mesh photonic crystals. Phys Rev Lett 76(14):2480–2483. https://doi.org/10.1103/PhysRevLett.76.2480
365. Vukusic P, Sambles JR (2003) Photonic structures in biology. Nature 424(6950):852–855. https://doi.org/10.1038/nature01941
366. Song M et al(2019). Colors with plasmonic nanostructures: a full-spectrum review. Appl Phys Rev 6(4): 041308. http://aip.scitation.org/doi/10.1063/1.5110051
367. Linden S et al (2006) Photonic metamaterials: magnetism at optical frequencies. IEEE J Select Top Quantum Electron 12(6):1097–1105. https://doi.org/10.1109/JSTQE.2006.880600
368. Cai W, Shalaev V (2010) Optical metamaterials. Springer, New York. https://doi.org/10.1007/978-1-4419-1151-3
369. Lapine M, Shadrivov IV, Kivshar YS (2014) Colloquium: nonlinear metamaterials. Rev Mod Phys 86(3):1093–1123. https://doi.org/10.1103/RevModPhys.86.1093
370. Urbas AM et al (2016) Roadmap on optical metamaterials. J Opt 18(9):093005. https://doi.org/10.1088/2040-8978/18/9/093005
371. Zhang L et al (2016) Advances in full control of electromagnetic waves with metasurfaces. Adv Opt Mater 4(6):818–833. https://doi.org/10.1002/adom.201500690
372. Chen H-T, Taylor AJ, Yu N (2016) A review of metasurfaces: physics and applications. Rep Prog Phys 79(7):076401. https://doi.org/10.1088/0034-4885/79/7/076401
373. Monticone F, Alu A (2017) Metamaterial, plasmonic and nanophotonic devices. Rep Prog Phys 80(3):036401. https://doi.org/10.1088/1361-6633/aa518f
374. Luo X (2018) Subwavelength optical engineering with metasurface waves. Adv Opt Mater 6(7):1701201. https://doi.org/10.1002/adma.201804680
375. Chang S, Guo X, Ni X (2018) Optical metasurfaces: progress and applications. Ann Rev Mater Res 48(1):279–302. https://doi.org/10.1146/annurev-matsci-070616-124220
376. Shaltout AM, Shalaev VM, Brongersma ML (2019) Spatiotemporal light control with active metasurfaces. Science 364(6441):1. https://doi.org/10.1126/science.aat3100
377. Wu Y et al (2006) Effective medium theory for magnetodielectric composites: beyond the long-wavelength limit. Phys Rev B 74(8):085111. https://doi.org/10.1103/PhysRevB.74.085111
378. Wang M, Pan N (2008) Predictions of effective physical properties of complex multiphase materials. Mater Sci Eng Rep 63(1):1–30. https://doi.org/10.1016/j.mser.2008.07.001
379. Markel VA (2016) Introduction to the Maxwell Garnett approximation: tutorial. J Opt Soc Ame A 33(7):1244. https://doi.org/10.1364/JOSAA.33.001244
380. Popov V, Lavrinenko AV, Novitsky A (2016) Operator approach to effective medium theory to overcome a breakdown of Maxwell Garnett approximation. Phys Rev B 94(8):085428. https://doi.org/10.1103/PhysRevB.94.085428
381. Lehmann J et al (2019) Microdisplays as a versatile tool for the optical simulation of crystal diffraction in the classroom. J Appl Crystallog 52(2):457–462. https://doi.org/10.1107/S1600576719001948
382. Kryuchkov M et al (2017) Antireflective nanocoatings for UV-sensation: the case of predatory owlfly insects. J Nanobiotech 15(1):52. https://doi.org/10.1186/s12951-017-0287-0

383. Kryuchkov M et al (2017) Alternative moth-eye nanostructures: antireflective properties and composition of dimpled corneal nanocoatings in silk-moth ancestors. J Nanobiotech 15(1):61. https://doi.org/10.1186/s12951-017-0297-y
384. Kryuchkov M et al (2020) Reverse and forward engineering of Drosophila corneal nanocoatings. Nature 585(7825):383–389. https://doi.org/10.1038/s41586-020-2707-9

Chapter 3
Experimental and Computational Methods

This chapter provides an overview of techniques that I have used for this work. After introducing the fabrication of artificial crystals, the major part discusses magnetic force microscopy as one of the two main techniques that have been applied for studying and manipulating the magnetic state of nanomagnetic arrays. The chapter continues by providing background for the second main part—optical methods, mainly the magneto-optical Kerr effect, that are able to detect changes in the magnetisation, but eventually of the toroidisation as well, exploiting more sophisticated experimental settings. Since the as-grown magnetic state of the fabricated crystals cannot simply be changed by annealing procedures, a concept for a non-thermal relaxation of the samples is presented to disprove a possible pinning of the magnetic configuration. A section of this chapter discusses a statistical analysis scheme to relate the local energy landscape to the formation of particular micromagnetic states. Moreover, a self-written two-dimensional micromagnetic calculation script is presented that gives access to magnetic fields that cause correlation between the building blocks in magneto-toroidal arrays. Last, the MFM-revealed microscopic magnetisation pattern do not show the corresponding toroidal order directly. Therefore, the chapter closes with ideas for uncovering contrast between different toroidal domain states using suitable image post-processing.
A new method that I have developed for teaching at ETH Zurich—a real time and interactive, optical simulation of crystal diffraction—is not covered in this chapter. Appendix A discusses its main aspects and motivates the published concept [1].

J. Lehmann, *Toroidal Order in Magnetic Metamaterials*, Springer Theses,
https://doi.org/10.1007/978-3-030-85495-9_3

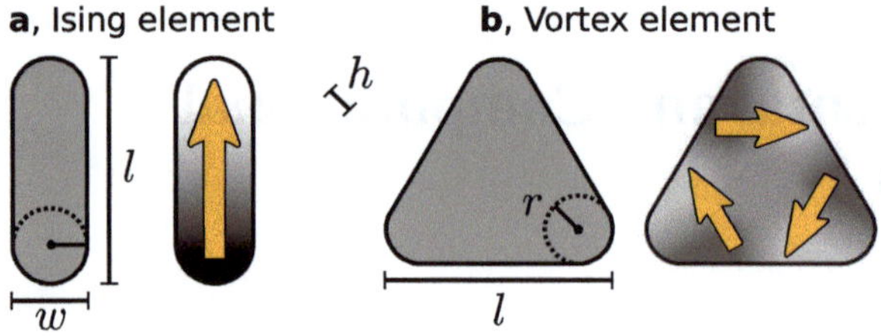

Fig. 3.1 Structural units of magneto-toroidal arrays. The arrays investigated during my work are based on Ising magnets **a** or vortex magnets **b** arranged in different tilings, see Fig. 4.1. Construction parameters such as the length l, width w and height h of the elements or curvature radii r quantify their geometry. Shown are the structural elements (grey) and visualisations of the internal magnetisation distributions (gradients and orange arrows that indicate the expected out-of-plane magnetic stray-field configurations) in the desired state. The arrows inside the triangle are chosen to indicate their connection with the resulting stray-field configuration

3.1 Fabrication of Nanomagnetic Arrays

Two main classes of structures have been grown, see Fig. 3.1, which differ in the character of fundamental building blocks.

Ising arrays: The fundamental building block represents a single giant spin and is formed by ferromagnetic material in a planar, stadium-like, elongated shape with an aspect ratio of $\geq 3{:}1$. This geometry ensures a shape-determined magnetic state with an in-plane Ising-like degree of freedom for each magnet, see Sect. 2.2 on page 24. Typical dimensions of my nanomagnets were around a few hundred nanometres by about one hundred nanometres. Structural cells were formed by 3, 4 or 6 nanomagnets, arranged as triangular, square or hexagonal crystals, respectively, with lattice constants in the range of $\leq 1\,\mu$m, see the overview in Fig. 4.1a–c on page 83.

Vortex arrays: The fundamental building block is a planar, equilateral triangle made from ferromagnetic material with rounded edges. The structure was chosen as a trade-off between two requirements. First, an internal magnetic vortex state had to form in each element due to weak in-plane anisotropy and a well-balanced magnetostatic and exchange energy. And second, a non-negligible remaining magnetic stray field was needed for an inter-element coupling of magnetic vortex states. A triangle's edge length of a few hundred nanometres with edge-curvature radii of ≈ 50 nm matched these design criteria. Crystal structures based on different tilings were realised using these building blocks, see the overview in Fig. 4.1d–f on page 83.
To maximise magnetostatic coupling, the magnetic elements were placed in close proximity to each other with separation distances from a few tens of to a few hundreds of nanometres. At the beginning of each experimental section within this book, corresponding geometrical specifications are provided. Investigated periodic arrays typically measure around $50 \times 50\,\mu\text{m}^2$ (only one set of samples used for optical-spectroscopy measurements has been grown as large as $3 \times 3\,\text{mm}^2$). For the fabri-

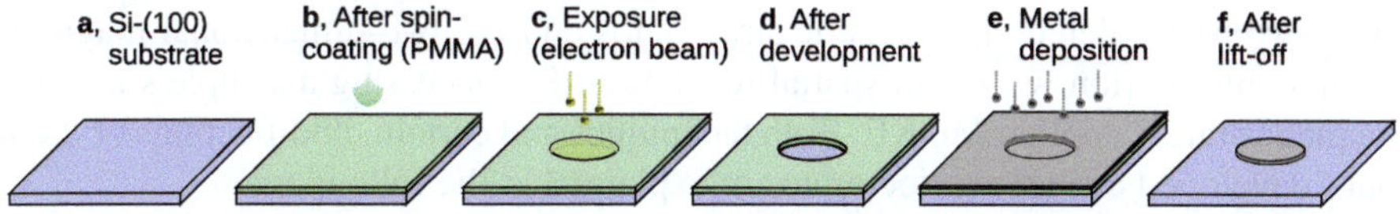

Fig. 3.2 Fabrication of the two-dimensional nanomagnetic structures. a A blank (100)-oriented silicon chip is used as a substrate. **b** The substrate is coated with a polymer resist. **c** Desired structures are exposed in the resist with an electron beam that locally alters the resist layer. **d** In the development step, the polymer layer can be dissolved in the exposed areas only. **e** Metal is deposited via an electron-beam assisted thermal evaporation. Two layers are deposited: a ferromagnetic metal (permalloy) of up to 30 nm and a capping layer (aluminium or gold) of 2–4 nm. **f** The final structure is obtained after a lift-off process that removes the unwanted resist and its covering metal layer

cation, we used a combination of electron-beam lithography, thermal evaporation deposition and lift-off at room temperature [2], see Fig. 3.2.

The samples were manufactured by Dr. Claire Donnelly and Dr. Naëmi Leo from the Laboratory for Mesoscopic Systems at the Paul Scherrer Institute. The desired structures have been patterned into a PMMA (Poly(methyl methacrylate)) layer coated onto a 200-μm or 500-μm-thick (100)-oriented silicon substrate using electron-beam lithography (Fig. 3.2a–c). After development, a negative of the structures was obtained (Fig. 3.2d). Then, permalloy ($Ni_{81}Fe_{19}$) as a ferromagnetic material was deposited via thermal evaporation with thicknesses up to 30 nm (Fig. 3.2e). The use of permalloy as a ferromagnetic material ensured a magnetic state determined primarily by shape anisotropy. The alloy provides a negligible magneto-crystalline anisotropy, see Sect. 2.2, and has been grown as a granular polycrystalline layer. A subsequent deposition of 2–4-nm-thick aluminium or gold prevented oxidation of the sample in air [3]. The lift-off (Fig. 3.2f), as the last step, removed the remaining resist and the unwanted material deposited on top of it.

3.2 Measurement of Microscale Magnetic Structures

Several methods exist that are capable of probing or imaging magnetisation patterns or spin-density modulations [4]. Some of these techniques, e.g. polarised neutron scattering (PNS) or x-ray-magnetic-circular-dichroism photoemission electron microscopy (XMCD-PEEM), require large-scale facilities such as spallation neutron sources or research reactors, or, synchrotron-radiation or free-electron laser beamlines, respectively. Established techniques on the laboratory scale include spin-polarised scanning-tunneling microscopy (SP-STM), MFM, MOKE, spin-resolved scanning electron microscopy (spin-SEM or SEMPA), Lorentz-electron microscopy, optical second-harmonic generation (SHG), superconducting quantum interference devices (SQUIDs) and others.

I used MFM for detecting magnetic stray fields and the two-dimensional magnetisation configuration with high spatial resolution. For accessing a sample's average magnetic state, I applied MOKE. Both techniques and certain modifications of them that I developed during my doctorate are explained in the following.

3.2.1 Magnetic Force Microscopy

MFM, as described in great detail in Refs. [4–10], is an experimental technique capable of imaging the spatial distribution of magnetic fields on the surface of a sample. It is used to study many varieties of magnetic structure with typically several tens of nanometres resolution lateral and sub-nanometre resolution of the height.

Measurement scheme

MFM was invented in 1987 [11, 12] as a subcategory of atomic force microscopy (AFM). As a variant of scanning probe microscopy (SPM), the technique is based on the interaction of a sharp tip at the end of a cantilever with the sample surface, see Fig. 3.3. Just like for AFM, the force between the tip and the sample is measured point-by-point in a two-dimensional xy-scan. Typically, a measurement is performed by tracing the deflection of laser light that is reflected from the top surface of the cantilever during the scan. Every local change in interaction strength $F(z)$ transfers to a variation in cantilever displacement or twisting and, thus, to a change in deflection of the laser light, which is monitored by a four-quadrant photodiode. The use of a segmented photodiode enables to differentiate bending- and buckling-like movements (vertical deflections) from torsional movements (horizontal deflections). Typically, the scan unit of the setup is based on piezoelectric actuators that can move the cantilever holder and/or the sample holder. A segmented piezo tube, for example, enables x-, y- and z-movements by an appropriately applied control voltage at individually addressable piezo segments. A piezo stack operates with three separate piezo elements that can elongate or contract along one of the Cartesian coordinate axes each. Usual scan ranges are of the order of up to $100 \times 100\,\mu\text{m}^2$. In addition, to control the motion of the cantilever, namely the height and its oscillation, another piezoelectric transducer is attached to the cantilever holder.

Tip-Sample Interactions

The core principle of all scanning probe techniques is to measure and possibly control the interaction strength $F(z)$ between the tip and the sample. Therefore, knowledge about its origin and functional dependence is crucial for the interpretation of measurement data. For a non-magnetic sample and tip and at short distances ($\approx$0–10 nm), the corresponding potential is defined mainly by the short-range attractive force that results from the overlap of electron orbitals (van-der-Waals interactions) and the repulsive force that is caused by Pauli exclusion within the atomic orbitals. Here I neglect contributions which are typically less pronounced, such as electrostatic forces from surface charges or capillary forces from condensed water films on sample and

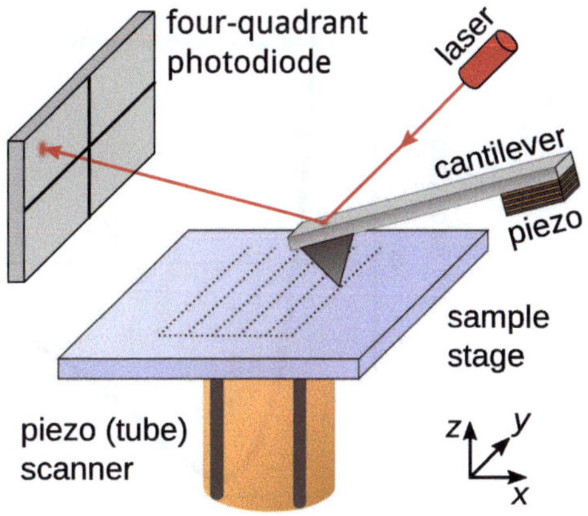

Fig. 3.3 Working principle of an atomic force microscope. The interaction strength between the tip at the end of the cantilever and the sample surface is mapped. The tip position is set by the scan unit (here, underneath the sample stage), while the cantilever motion is controlled by the piezoelectric actuator attached to the sample holder. The interaction-strength measurement is performed optically by monitoring the position of a reflected laser beam with a segmented photodiode

tip in air. The resulting force can be approximated by the Buckingham potential $V_{\mathrm{B}}(z)$ [13],

$$V_{\mathrm{B}}(z) = P \cdot \exp\left(\frac{-z}{z_0}\right) - \frac{W}{z^6} \quad , \tag{3.1}$$

where z is the distance between tip apex and sample surface, while the parameters P and z_0 correspond to the repulsion due to Pauli exclusion and the parameter W quantises the van-der-Waals interaction. The total force $F(z)$ results from the negative derivative of the potential as

$$F(z) = -\frac{\partial V_{\mathrm{B}}(z)}{\partial z} = \frac{P}{z_0} \cdot \exp\left(\frac{-z}{z_0}\right) - \frac{6W}{z^7} \quad , \tag{3.2}$$

and is shown in Fig. 3.4. Based on this function, different measurement modes to sense the interaction strength can be realised.

Tapping or Semi-contact Mode

One of the major modes of operation within an atomic force microscope is the tapping mode [14] which is a mode that traces the local minimum of the force curve. Here, the cantilever is excited at its resonance frequency, typically in the range of 100 kHz, using the piezoelectric transducer attached to it. Starting from the equilibrium position, the resonance frequency of the cantilever will be blue-shifted by any repulsive force (tip is approaching the sample) or red-shifted by an attractive force (tip detaches from the sample). When the excitation frequency is fixed at the free resonance of the cantilever, any red- or blue-shifts will transform to changes

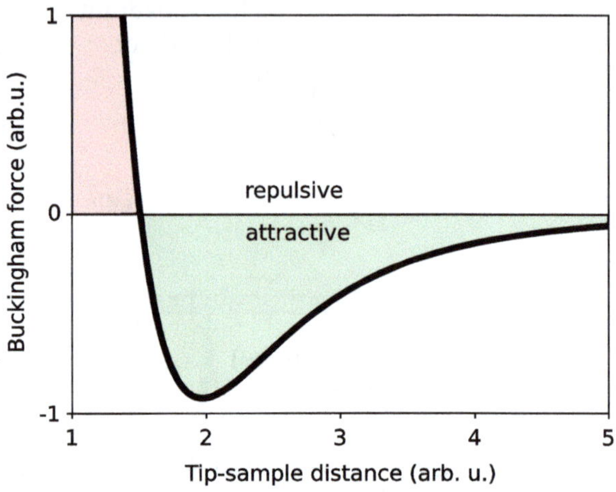

Fig. 3.4 Force between tip apex and sample surface. Due to competition between attractive van-der-Waals forces and repulsive Pauli-exclusion forces, the system shows a minimum at a finite distance z. Here, the plotted force curve is based on the Buckingham potential [13] following Eq. 3.2. The red area indicates repulsive forces, while the green area indicates attractive forces

in the resonance amplitude. Since both a blue- or a red-shift result in a decrease in amplitude, attractive or repulsive forces can not be separated. Fortunately, the change in resonance frequency also transforms into phase shifts $\Delta\phi$ of the optically detected signal relative to the electrical excitation signal with opposite sign for attractive or repulsive forces as [6, 15]

$$\Delta\phi \approx \frac{Q}{k}\frac{dF(z)}{dz} , \tag{3.3}$$

where Q is the quality factor of the mechanical cantilever oscillation and k is the cantilevers spring constant. This phase shift can be used as a correction signal to the height-controlling piezo actuator which, hence, enables the tip to hover across the scanned area with a constant interaction strength. By monitoring the regulation voltage of the piezo for the z-coordinate, the topography of the sample can be mapped. The maximum speed for a two-dimensional scan is restricted by the accuracy of the feedback loop and, therefore, determined, among others, by the time constant of the oscillating cantilever (given by its mechanical damping factor). Mechanical vibrations of the instrument, thermal noise of the cantilever and electrical noise within the optical detection system limit the achievable sensitivity along the z-direction. The resolution in xy-direction is limited primarily by the tip-apex radius which is typically of the order of 50 nm for commercially available MFM tips.

Measurement of Magnetic Samples

For MFM as compared to AFM, the tip is coated with a ferromagnetic layer, while the sample is typically a material that displays a magnetic order.[1] As a result, an additional energy term has to be taken into account that represents the magnetostatic

[1] This might directly originate from ferro- or ferrimagnetism, or, for instance, from non-compensated magnetic moments at domain walls in antiferromagnets or from induced spin separations and accumulations.

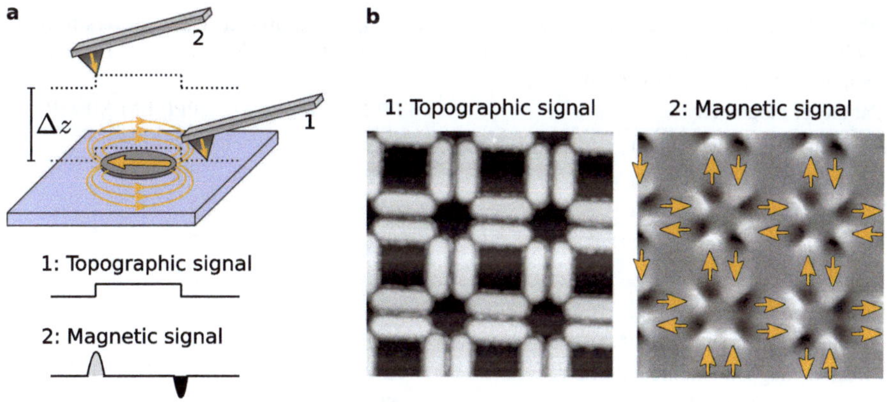

Fig. 3.5 Magnetic force microscopy in the lift-height or double-pass mode. a In the first pass, the system probes short-range forces to map the samples topography. In the second pass, the tip is lifted by Δz of typically tens of nanometres and retraces the topography. Now, long-range forces that result from the magnetostatic interaction determine the measured signal and allow to map the volume and surface magnetic-charge distribution of the sample. Sketched is an exemplary measurement (single line scan) of a single-domain in-plane-magnetised ferromagnetic element. **b** MFM measurement of a section of an array composed of in-plane magnetised nanomagnets

interaction (see Sect. 2.2.4 on page 35) between the tip and the sample. The spins in the thin ferromagnetic layer deposited on an MFM tip experience a strong shape anisotropy, see Sect. 2.2.1, that forces the net magnetisation to align parallel to the symmetry axis of the tip. The application of a strong magnetic field along the tip axis ensures a single-domain state within the coating. For simplicity, it is common to model the magnetic moment of the tip apex as a homogeneously magnetised prolate spheroid or by just a point-like magnetic dipole [16, 17]. Since the decay of magnetic-dipole fields shows a distance dependence of z^{-3}, it dominates at farther distances over the local interaction-energy terms approximated in Eq. 3.1. Therefore, the contributions due to magnetic order can be disentangled simply by using a two-pass scanning mode, see Fig. 3.5.

The first pass in tapping-mode senses the interactions that are the strongest at very short distances ($\lesssim 10$ nm) as given by Eq. 3.1. The resulting xy-map represents the sample's height profile in the so-called topography-channel of the instrument. After each line, the tip retraces the contour of the first pass at a desired lift-height Δz of typically tens of nanometres. This second pass maintains a constant tip-sample distance while sensing longer-distance decaying forces such as the magnetic dipole-dipole interaction. The lift-height Δz is chosen to be large enough to avoid any interference of topographical signatures in the second pass, whereas an upper limit results from the decreasing magnetostatic interaction which is accompanied by a decreasing signal strength at too large distances. The signal extracted in the second pass corresponds primarily to the out-of-plane magnetic-field component caused by surface or volume magnetic charges, see Sects. 2.2.1 and 2.2.1. In particular for the nanomagnetic elements investigated in this work, surface magnetic charges are

Table 3.1 Cantilevers used for magnetic force microscopy measurements. Cantilevers from the supplier NANOSENSORS [18] have been used for imaging (PPP- LM- MFMR) as well as for magnetic manipulation (PPP- MFMR) of the samples

Cantilever parameter	PPP-MFMR	PPP-LM-MFMR
Resonance frequency (kHz)	75	75
Force constant (N/m)	2.8	2.8
Thickness (μm)	3	3
Mean width (μm)	28	28
Length (μm)	225	225
Coercivity (mT)	30	25
Remanence magnetisation (A/m)	3×10^5	1.5×10^5
Effective magnetic moment (A m^2)	$(10 - 13) \times 10^{-3}$	$(5 - 7) \times 10^{-3}$
Magnetic resolution (nm)	<50	<35

detected as positive or negative phase shifts in the optically detected eigenfrequency of the cantilever. This signal can be transferred to a black-and-white contrast in corresponding images.

Instrumentation

All MFM measurements presented in this book have been performed on the NT-MDT SI NTEGRA PRIMA [18]. The scanning-probe system is a modular and flexible platform that allows for very versatile experimental settings, such as the probing of heavy and bulky samples, large scan areas (up to $100 \times 100\,\mu\text{m}^2$) or measurements in different atmospheres. In addition, the scanning, positioning and detection unit is made from non-ferromagnetic materials. This, together with the in-plane as well as out-of-plane magnetic-field coils provided by NT-MDT (generating magnetic flux densities of $\approx 100\,\text{mT}$), offers capabilities for a magnetic imaging during the application of magnetic fields. To avoid measurement artefacts and to ensure a linearly responding device, each of the x, y and z piezo scan elements in the SMENA measurement head is equipped with a capacitance-measurement-based closed-loop control.

The specifications for the MFM tips used throughout the presented work are listed in Table 3.1. These tips are coated with a 40-nm-thick cobalt-alloy layer, providing a permanent magnetisation along the tip axis which can be set using a strong $Nd_2Fe_{14}B$ permanent magnet. All scans for imaging have been performed using the NANOSENSORS PPP- LM- MFMR cantilevers, whereas the NANOSENSORS PPP- MFMR have been used for micromagnetic manipulation.

3.2.2 *Magneto-Optical Kerr Effect*

MOKE, discovered 1877 by John Kerr [19, 20], is a phenomenon of light-matter interaction that appears as a modification in the polarisation state of a light beam upon its reflection from a magnetically ordered surface. The resulting effect is a rotation of the polarisation by θ_K and a change in ellipticity δ_K, defined as the amplitude ratio between the major and the minor axis. While the following section gives a brief introduction and insight into the underlying physics of MOKE, a more detailed and comprehensive review can be found in Refs. [21–26].

Formally, the phenomenon can be described as a combination of magnetic-circular birefringence (different phase velocities for the two circular-polarised components of a light field) and magnetic-circular dichroism (different absorption coefficients for the two circular-polarised components of a light field), which implies an inequivalent interaction of left- and right-handed circular polarised light with the magnetisation of a material. Mathematically, the interaction can be written in a general expression ($\hat{\epsilon}$) of the dielectric permittivity tensor $\epsilon = \epsilon_0 \epsilon_r$ that relates the dielectric displacement $\vec{D}$ in a material with the electric field $\vec{E}$ of a light wave,

$$\vec{D} = \hat{\epsilon} \vec{E} \ . \tag{3.4}$$

This reformulation adds the magnetically induced circular birefringence as a Lorentz-force-like *gyroelectric* term as

$$\vec{D} = \epsilon (\vec{E} - \vec{E} \times \mathrm{i} Q \vec{m}) \ , \tag{3.5}$$

since the symmetry of the interactions phenomenologically relates to a Lorentz force acting on the electrons that are accelerated by the light's electric field. Consequently, the expanded form of this tensor reads for isotropic media or for cubic crystals as

$$\hat{\epsilon} = \epsilon \begin{pmatrix} 1 & -\mathrm{i}Qm_z & \mathrm{i}Qm_y \\ \mathrm{i}Qm_z & 1 & -\mathrm{i}Qm_x \\ -\mathrm{i}Qm_y & \mathrm{i}Qm_x & 1 \end{pmatrix} \ , \tag{3.6}$$

where Q is the complex, frequency- and material dependent magneto-optical coupling constant (so-called Voigt constant) and $m_i = M_i / M_S$ are the magnetisation components M_i in Cartesian coordinates, normalised to the saturation magnetisation M_S. The interaction is, thus, linear in the magnetisation and depends primarily on the relative orientation of the light's wave vector κ and the direction of magnetisation $\vec{m}$ relative to the sample surface.
Three experimental settings are distinguished: The polar, the longitudinal, and the transversal MOKE geometry, see Fig. 3.6. Since only in-plane magnetised samples are used during my studies, only the longitudinal (LMOKE) or the transversal (TMOKE) are applicable. Following the trajectory of a Lorentz-force-induced

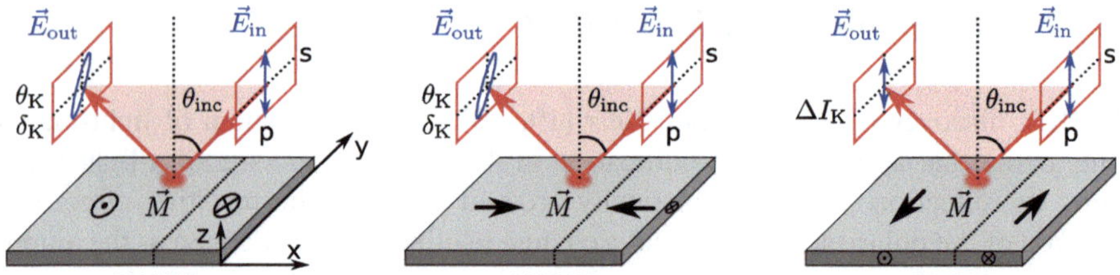

Fig. 3.6 Basic geometries for measuring the magneto-optical Kerr effect. The experimental settings for probing the magneto-optical Kerr effect are classified in three groups depending on the direction of magnetisation (black arrows) with respect to the sample surface and plane of incidence. In the polar geometry (PMOKE, left), the magnetisation $\vec{M}$ points along the surface normal (z-direction) and lies in the plane of incidence (light red plane). The Kerr rotation θ_K and ellipticity δ_K scales with cos (θ_{inc}) and is nearly independent on the direction of input polarisation $\vec{E}_{in}$. For the longitudinal geometry (LMOKE, centre), an in-plane magnetisation lies in the plane of incidence. The polarisation rotation θ_K and ellipticity δ_K scales with sin (θ_{inc}) and applies for a polarisation parallel to the plane of incident (p) as well as perpendicular to it (s). For the transversal geometry (TMOKE, right), an in-plane magnetisation points perpendicular to the plane of incidence. Instead of a Kerr rotation, the signal is a change in intensity ΔI_K for an input-polarisation vector that is parallel to the plane of incidence (p). The classification of an experiment into precisely one of the three geometries is not generally applicable due to a potential intermixing of contributions from the different MOKE geometries [27]

motion of itinerant electrons, LMOKE yields an equally strong Kerr rotation but with opposite sign for both orthogonal polarisation components, i.e., parallel and perpendicular to the plane of incidence (p and s polarisation, respectively). The polarisation rotation, and thus the amplitude of the signal, scales with $\sin(\theta_{inc})$, where θ_{inc} is the angle of incidence (see Fig. 3.6). Hence, the signal is at maximum for large angles of incidence and equal zero for normal incidence since here the Lorentz force points parallel to κ and therefore does not excite any radiating dipole.[2] In contrast, the orientation of the Lorentz-force vector for the transverse geometry with an incident electric-field vector E_{inc} parallel to the plane of incidence implies the generation of a radiating component with a polarisation that points parallel to the incident-light polarisation and scales with $\sin(\theta_{inc})$. Thus, the interference of these two terms results in an amplitude change for the reflected light. Since, typically, the measurement of a change in polarisation can be done with higher accuracy than the measurement of a change in intensity of a light beam, LMOKE is the preferred mode of operation throughout this work and will be discussed together with experimental details in the following section.

In a microscopic formalism, the magneto-optical Kerr effect originates from Zeeman splitting of orbitals either induced by an external field or due to the presence of an internal magnetic order [28–31]. These orbitals shift in energy due to spin-orbit interaction (Therefore, a large Voigt constant can be expected from materials with

[2] Note that, due to the large refractive index of conductors, light refracts into a metal with almost normal incidence. The consequence is a generally smaller signal for LMOKE than for PMOKE.

strong spin-orbit interaction [24]) and, as a consequence, modify the dispersion relation for light with different helicities. In turn, the modified dispersion leads to different refractive indices and different absorption coefficients for left- or right-handed circular-polarised light and, thus, to a change in the light's polarisation state after the interaction.
The penetration depth of light into a metal is typically a few tens of nanometres It determines the depth of information [32] of about 10 nm that is accessible with MOKE. Accordingly, MOKE is a very surface sensitive technique, suitable for the study of layered systems [33, 34] or—as investigated here—patterned thin films.

Setup for Optical Magnetisation Measurements

Magnetisation reversal curves from arrays of nanomagnetic Ising- or vortex-type elements have been recorded via polarisation analysis within a self-built computer-controlled MOKE setup, see Fig. 3.7. Two continuous-wave laser-light sources of different wavelengths λ have been used, since laser, as compared with conventional incoherent light sources, offer the advantage of high spectral intensity combined with very stable output power. On the one hand, a helium-neon laser emitting an optical power of about 2 mW at a wavelength of 633 nm (photon energy of 1.96 eV) has been used due to its high stability and excellent beam profile. A simple Galilean telescope allows for an expansion of the otherwise very narrow laser-beam diameter. On the other hand an actively current-stabilised laser diode emitting a power of up to 100 mW at a wavelength of 405 nm (photon energy of 3.06 eV) has been used.

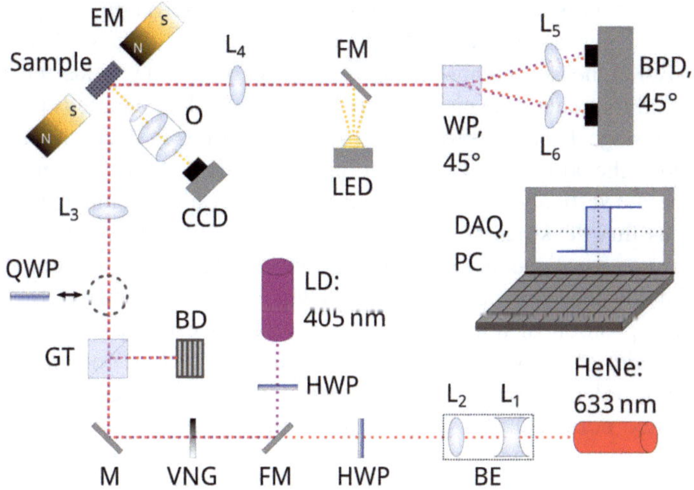

Fig. 3.7 Experimental setup for magneto-optical measurements. Abbreviations—LD: laser diode, HeNe: Helium-Neon laser, L_i: ith lens, BE: beam expander, HWP: half-wave plate, FM: flip mirror, VNG: variable neutral grey filter, M: mirror, GT: Glan-Taylor prism, BD: beam dump, QWP: quarter-wave plate, EM: electromagnet, O: 10× objective, CCD: camera, LED: light-emitting diode, WP: Wollaston prism, BPD: balanced photo diode, DAQ, PC: computer-based data acquisition system

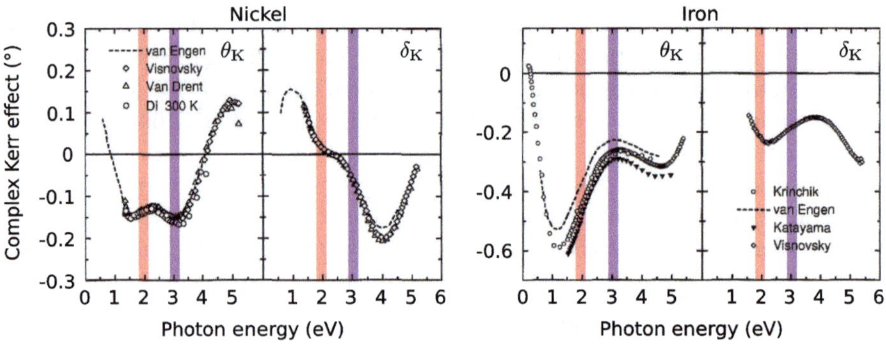

Fig. 3.8 Magneto-optical response of nickel and iron as a function of the photon energy. Spectra of the Kerr rotation (θ_K) and ellipticity (δ_K) from the infrared to the ultraviolet spectral region. The positions of two laser photon energies used for my work are marked in red (1.96 eV) and violet (3.06 eV), respectively. Figure adapted from Ref. [38]

Both laser wavelengths address the broad spectral region of strong Kerr rotation $|\theta_K| > 0.1°$ in nickel, iron and their alloy [22, 31, 35–39], which lies in between photon energies of about 1 eV and 3.5 eV, see Fig. 3.8.

The polarisation of the light emitted by the laser sources can be rotated by half-wave plates and a variable neutral-grey filter is used to set the initial intensity. A Glan-Taylor prism subsequently defines the final polarisation state of the light beam before it is focussed under an incident angle of $\theta_{inc} = 45°$ onto the sample by a lens with 100 mm focal distance. As a result, the spot diameter on the sample surface is around 30 μm which is slightly smaller than the typical size of the studied arrays. For advanced MOKE measurement schemes, see the section below, a quarter-wave plate can be inserted for generating circularly polarised light. The sample is mounted on a non-magnetic stage that allows for x, y, z, θ, and ϕ degrees of freedom. Furthermore, it is mounted in the bore of a self-made computer-controlled four-axis electromagnet that can generate vertically, diagonally and horizontally aligned magnetic fields with homogeneous flux densities of $B \approx 250$ mT inside its almost 20-mm-sized square-shaped aperture, see Appendix B.2. In addition, the sample can be mounted onto a simple self-made out-of-plane electromagnet which provides a magnetic-flux density at the sample position of about 200 mT. Since the investigated samples typically consist of a series of arrays deposited on a single silicon chip, imaging capabilities are required in order to locate a particular array of interest. Therefore, an objective lens is mounted normal to the sample surface that magnifies and brings an image of the sample onto a camera sensor. Illumination is performed with an orange light source that can be guided into the beam path from the detection side, meaning that the camera can image the illuminated sample together with the incident laser beam, allowing for a fine adjustment of laser focus and lateral position. After the light has interacted with the sample, a second lens with 100 mm focal length re-collimates the beam and guides it to a Wollaston prism, which is mounted with its main axis under 45°. The prism splits the beam into two orthogonal polarisation states, which have equal intensity (in case of a reflection from a non-magnetic sample such as the silicon substrate) and leave the prism with a symmetric opening angle of 18°. Subsequently, lenses with a focal distance of 75 mm focus the two beams onto the two silicon diodes of a self-made balanced photo detector, see Appendix B.3. The advantage of such a detection system lies in proportionality between the output voltage of the detector and the intensity difference measured on the two diodes. An equal intensity in both beams results in zero signal, while every imbalance, here, due to a magneto-optical interaction, results in an output voltage with negligible sensitivity to, e.g., amplitude noise or fluctuations. The setup as a whole is shielded from surrounding light and, in addition, designed with the idea in mind that the best signal-to-noise ratio can be achieved by avoiding that the probing light beam passes unnecessary optical components such as (dielectric) beam splitters, coated objective lenses or similar that may alter the magnetic information encoded in the light's polarisation state.

Diffraction MOKE and Differential Diffraction MOKE

Diffraction MOKE (DMOKE), also called Bragg-MOKE or non-specular MOKE, was first observed in 1963 [40] as an optical diffraction phenomenon related to the presence of a 'ferromagnetic grating' [41, 42]. Since then, the technique has proven to provide insight into re-magnetisation processes and micromagnetic configurations of periodic magnetic structures such as naturally occurring periodic domain pattern[3] [40, 42–45] as well as synthetic one- or two-dimensional grating-like structures [7, 46–58]. While here, only a brief outline of the concept is given, an in-depth review or extensive application can be found in Refs. [59–62].

DMOKE is based on the idea that periodic magnetic structures cause optical diffraction such that a MOKE analysis is not restricted to the specularly reflected beam, but can be performed 'off-specular' in higher orders n of diffraction. The idea benefits from k-space resolution, similar as in structural-refinement crystallography (see also Sect. A.1), such that magnetic information on the level of a unit cell can be extracted from a non-specular-MOKE-based hysteresis loop or, e.g., a magnetic dark-field imaging can be performed using diffracted light. In analogy to atomic form factors in an x-ray diffraction experiment, magnetic surface form factors can be calculated by an integration over the spatial (position vector $\vec{r}$) magnetisation distribution $\vec{m}(\vec{r})$ of a unit cell (uc) with $\vec{G}$ as a reciprocal lattice vector as

$$f_n = \int_{\text{uc}} \vec{m}(\vec{r}) \exp\left(\mathrm{i}\, n\, \vec{G} \cdot \vec{r}\right) d^2 r \quad . \tag{3.7}$$

Typical for DMOKE is that a symmetric scattering geometry is assumed, meaning that monitoring of only positive (or only negative) orders of diffraction is sufficient to extricate the full information accessible by the experiment.

[3] The domain states do not necessarily have to form a strictly periodic alternating pattern, but should have a characteristic length scale of the order of the light wavelength.

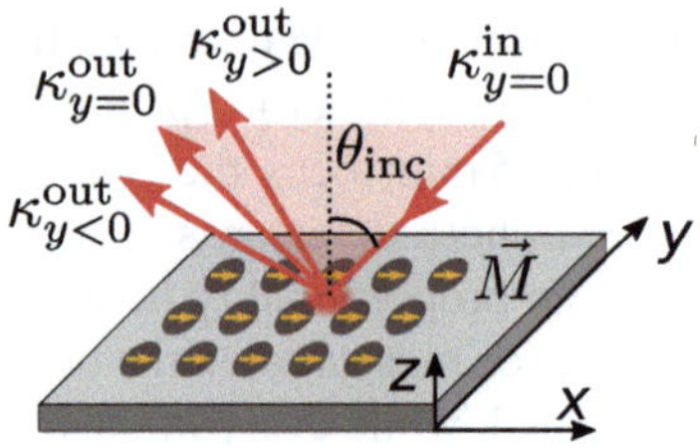

Fig. 3.9 Principle of differential diffraction MOKE. The laser-light beam interacting with the sample surface under an incident angle of $\theta_{inc} = 45°$ (xz-plane) experiences diffraction due to the presence of a structural or magnetic (orange arrows) grating. While the incident light beam lies in the xz-plane (wave vector component $k_y = 0$), diffraction spots may appear with either positive or negative y-values. Thus, the intensity difference of these two diffraction spots can encode a directional anisotropy

This is where differential DMOKE (ΔDMOKE), a novel technique that I developed during my doctorate, goes one step further: The new method uses non-specular MOKE to achieve a sensitivity to non-reciprocal directional anisotropies (NDA), which implies that a material property does not remain invariant under the exchange of source and detector in an experiment [63]. Such an anisotropy may result from a magneto-chiral effect or an optical magnetoelectric effect that occur in certain materials that violate both space-inversion and time-reversal symmetry [64–70], as, for example, ferrotoroidic materials, see Sect. 2.1.5.
The technique is based on the simultaneous measurement of the intensity (or polarisation rotation) in two diffraction spots that both carry magneto-optical signatures and, furthermore, have the same diffraction order, yet with an opposite sign. The wave vectors of these scattered light beams $\vec{k}_{\pm n}$ encode directional anisotropies induced by the sample. Hence, this measurement principle can circumvent the need for changing the light path or the orientation of the sample during the experiment. However, it requires a flexible detector arrangement, since, unlike in typical DMOKE setups, see Ref. [62] p. 74, the two chosen diffraction spots leave the sample neither parallel to the plane of incidence nor at well-defined angles, see Fig. 3.9.

3.3 Demagnetisation—A Non-thermal Relaxation Protocol

In order to investigate relaxation behaviour, spontaneous domain formation and to test for memory effects and domain pinning by topographic defects, demagnetisation protocols have been applied to nanomagnetic arrays. The experiments address the following questions:

- Can a demagnetisation protocol relax a toroidal array into its single-domain ground state?
- Are domain walls pinned by defects and therefore coupled to structural inhomogeneities?

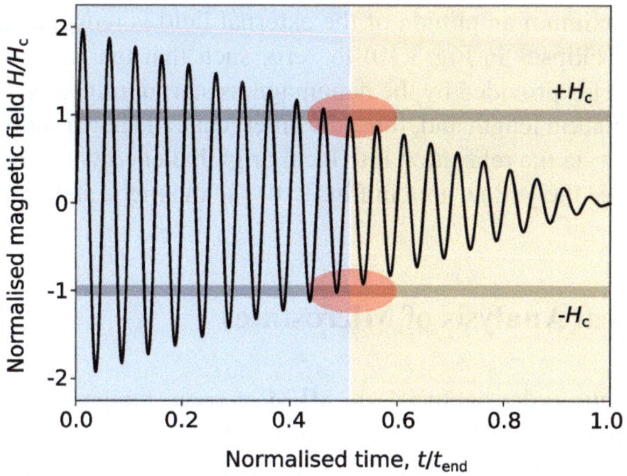

Fig. 3.10 Principle of AC-field demagnetisation. The sample is exposed to an oscillating magnetic field H superimposed with a monotonous decreasing amplitude over time t. The Initial amplitude is set to be well above the samples coercive field $|H_c|$ (horizontal grey lines). Accordingly, the magnetisation of the sample follows the applied magnetic field (blue shading). The external field amplitude is decreased (here, linearly). At the time, the external magnetic field is of comparable amplitude to the magnetostatic fields within the sample (the coercive field H_c in, for instance, a nanomagnetic array), the correlations between neighbouring magnets induce ordered states (red shading). Finally, a relaxed state becomes frozen as the external field is swept to zero (orange shading) at t_{end}

The so-called AC-field demagnetisation is a treatment of magnetic samples with external field sweeps to mimic thermally induced spin-flip events typically found during an annealing or growth procedure [71, 72]. In the context of nanomagnetic arrays, which are the main sample systems for my work, brief introductions into the concept and application examples are given in Refs. [73–75]. AC-field demagnetisation is specifically interesting for the case of nanomagnetic arrays, since typical blocking temperatures (see Sect. 2.2.2 from page 30 on) of >10-nm-thick macrospins are in the range of several 10^4 K [76].
The demagnetisation is experimentally done by applying a magnetic field $\vec{H}$ to a sample that—in the beginning—overcome the sample's coercive field: $|\vec{H}| > |\vec{H}_c|$. At this stage, the direction of magnetisation prefers an alignment towards the external magnetic field. Now, by continuously changing the orientation of the external field, the direction of magnetisation of single nanomagnetic elements flips while following the external field direction. Macroscopically, this can be seen as an analogy to the superparamagnetic phase, see Sect. 2.2.2, at which the magnetisation in single-domain nanomagnets are subject to random reorientations due to the thermal energy. Here, however, these reorientation events are deterministically driven by the rapidly alternating direction of the external field, see the blue area in Fig. 3.10. For structures with an in-plane magnetisation, as investigated here, it is sufficient to only sweep the external field direction in the plane to access a large number of (stable and metastable) local magnetic configurations. In the second and most relevant stage of the

protocol, the maximum amplitude of the external field is reduced steadily through H_c (see the red ellipses in Fig. 3.10) to zero, such that the influence of the local magnetostatic fields provided by the nanomagnets start dominating. The result is an increasing correlation length and, thus, the emergence of short-range and eventually long-range order via the relaxation into local or global energy minima, respectively. Far below H_c (see the yellow area in Fig. 3.10) the magnetic state remains frozen.

3.4 Statistical Analysis of Microstates

Since the magnetic order captured via MFM at room temperature corresponds to a 'high-temperature' equilibrium state that has become frozen during the growth process [71, 72], a statistical analysis of the local order is applicable. Such a statistical treatment connects calculated energy levels for different configurations of local ensembles of spins to their experimentally determined probabilities and vice versa. Hence, a Boltzmann analysis allows one to trace the occupation density of particular types of excited states (e.g. domain-wall states in the mesoscopic nanomagnetic arrays). This theoretical handle can be very useful for the control of an emergent order and for (im)balancing different low-energy excited states that finally determine the macroscopic behaviour of the studied system.

The primary coupling promoting magnetic order in the investigated systems is the magnetic-dipole interaction, as introduced in Sect. 2.2.4 on page 35. Because dipole fields decay quickly with $|\vec{r}|^{-3}$, only nearest neighbours can be considered in the first approximation.[4] This allows for a simple formal description of local order in the Ising-type toroidal square array by considering only the magnetic configurations of subunits consisting of eight Ising-spins that meet at a so-called vertex, see Fig. 3.11. The magnetic-dipole energy per vertex is composed of perpendicular ($E_\perp$) and parallel ($E_\parallel$) contributions that directly represent the two distinct terms in the dipole-interaction energy, see Eq. 2.18,

$$E_\perp = \frac{\mu_0}{4\pi}\left(\frac{-3\,(\vec{m}_i\,(\vec{r}_i-\vec{r}_j))\,(\vec{m}_j\,(\vec{r}_i-\vec{r}_j))}{|\vec{r}_i-\vec{r}_j|^5}\right) \quad ; \quad E_\parallel = \frac{\mu_0}{4\pi}\left(\frac{\vec{m}_i\,\vec{m}_j}{|\vec{r}_i-\vec{r}_j|^3}\right). \tag{3.8}$$

For the Ising-type toroidal square array, this leads to the following total energy per vertex:

$$E_{\text{vertex}} = n_\parallel\,E_\parallel + n_\perp\,E_\perp \quad ; \quad n_i \in \{1, 2, 3, 4\} \ . \tag{3.9}$$

[4] It can be shown that two-dimensional magnetic structures exhibit a much weaker correlation as compared with three-dimensional ones. The integration of the dipole-interaction potential ($\propto r^{-3}$) over a three-dimensional volume of interacting magnetic moments yields $\lim_{r'\to\infty}\int_0^{r'} r^{-3}\cdot r^2 dr = \ln(r') + \text{const.}$ and, thus, diverges for $r'\to\infty$. In the case of a two-dimensional system, the area-integral yields $\lim_{r'\to\infty}\int_0^{r'} r^{-3}\cdot r dr = -1/r' + \text{const.}$, which gives finite value for $r'\to\infty$.

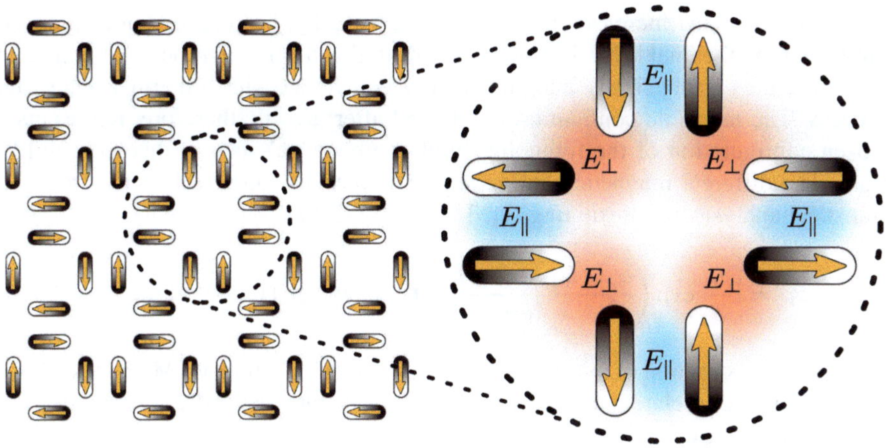

Fig. 3.11 Parametrising the interactions in toroidal square arrays. Left: The local couplings can be captured by considering a single vertex at which eight macrospins meet. Right: Such a vertex is stabilised by the interaction-energy terms $E_{\perp}$ (red) and $E_{\|}$ (blue). The vertex energy can be taken as the sum over the energy contribution from each of the local couplings, see Eq. 3.9

Here, n_i is defined as the number of head-to-head or tail-to-tail alignments of neighbouring magnetic moments. Equations 2.18 and 3.9 leads to positive and negative values of equal magnitude for head-to-head/tail-to-tail and head-to-tail alignments, respectively. It is, however, reasonable to define the energy for a head-to-tail configuration to be zero such that only unfavoured constellations contribute to the total vertex energy. The full diagram of vertex-energy states (see Fig. 5.6 on page 97) includes $2^8 = 256$ different magnetic configurations.
Now, the probability P_i to observe a particular (i-th) microstate in a thermally active system is directly attributed to the microstate's energy E_i, see the Boltzmann-distribution Eq. 2.3 in Sect. 2.1.2. In my study, systems are usually thermally inactive at room temperature and appear highly correlated. Yet, a statistical treatment is eligible for describing systems that have become frozen during their thermally active path towards order. During the growth of ferromagnetic Ising-type nanostructures, see Sect. 3.1, each macrospin passes the superparamagnetic phase at which the thermal energy overcomes the switching barrier and induces fast random reorientations in the magnetisation between the two Ising states, see Sect. 2.2.2. This behaviour typically occurs at a few nanometres thickness of the deposited ferromagnetic structures [71, 72]. In this stage of the growth, the correlation length increases with thickness and the system localises its local or global energy minimum at finite temperatures.[5] At a later stage of the deposition of ferromagnetic material, the mag-

[5] Note that, for a quench from temperatures that correspond to a thermal energy of a magnitude that is comparable to the magnetic coupling energy, entropy provides a non-negligible contribution to the formation of short-range order. This in turn is the reason for the thermal population of vertex states of different energy.

netic states become increasingly stable and, eventually, become frozen or blocked until reaching the final layer thickness. Accordingly, the growth process of magnetic nanostructures is physically and statistically equivalent with a thermal annealing process. The magnetic configuration obtained after growth, therefore, represents a frozen configuration of the foregoing 'high-temperature' state in thermal equilibrium. The different magnetic configurations of vertices can now be subjected to a statistical analysis by substituting Eq. 3.9 into Eq. 2.3 as

$$P_i = \frac{g_i}{Z} \exp\left(-\frac{n_\parallel E_\parallel + n_\perp E_\perp}{k_B T}\right) \quad ; \quad n_i \in \{1, 2, 3, 4\} \ . \tag{3.10}$$

With this, an extraction of microstate occupation densities from MFM images allows to experimentally relate local order to the energy values $E_\perp$ and $E_\parallel$, normalised to the thermal energy.

3.5 Calculations of Short-Range Magnetostatic Couplings

To reveal the configuration of the magnetic field that is generated by an ensemble of Ising-type nanomagnets of defined magnetic state and to calculate its possibly demagnetising effect on neighbouring macrospins, numerical micromagnetic calculations have been implemented in a flexible self-written PYTHON script.

The idea behind the script is to calculate point-like or laterally averaged magnetic fields that are generated from a desired number of magnets placed in the surrounding, with a defined position and an Ising-like magnetised state. The script first defines a two-dimensional square-celled mesh as a basis for the calculation. A cell size of $5 \times 5\,\text{nm}^2$ is chosen, on the one hand, to match the magnetic exchange length of permalloy [77], such that the assumption of a homogeneous magnetisation within a single cell is supported, and, on the other hand, that the edges of magnetic elements (see Fig. 3.1) can be smoothly approximated. Now, building blocks have been defined as zero-dimensional (macrospin approximation), one-dimensional (magnetic needle approximation) or two-dimensional (no further approximation, yet a homogeneous and discrete magnetisation assumed) areas, consisting of 1, 91, or 2585 homogeneously magnetised cells, respectively (see blue dots in Fig. 3.12). For the approximations, the total magnetic moment of a macrospin, see Sect. 2.2.2, is distributed evenly between the corresponding cells and the magnetic-dipole field of all cells is superimposed to yield the resulting global field distribution. Quantitatively, the magnetic-dipole field $\vec{H}_D(\vec{\mu}, \vec{r})$ generated by one cell with magnetic moment $\vec{\mu}$ located at $\vec{r}_0$ is calculated as:

$$\vec{H}_D(\vec{\mu}, \vec{r}) = \frac{1}{4\pi} \left(3\, \frac{\vec{\mu}\,(\vec{r} - \vec{r}_0)}{|(\vec{r} - \vec{r}_0)|^5}\,(\vec{r} - \vec{r}_0) - \frac{\vec{\mu}}{|(\vec{r} - \vec{r}_0)|^3}\right) \ . \tag{3.11}$$

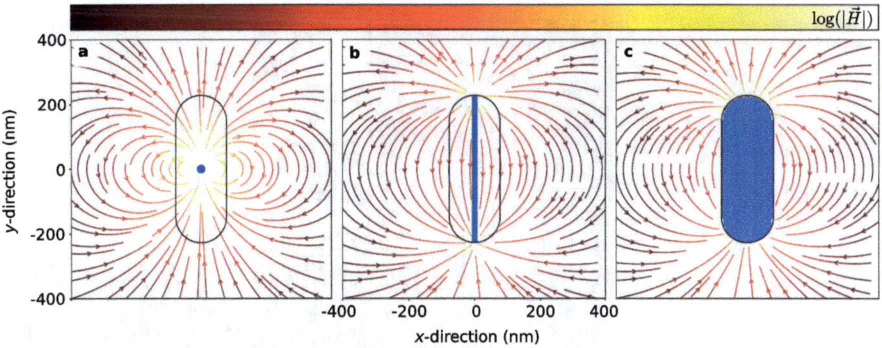

Fig. 3.12 Magnetic field configurations for different nanomagnet approximations. The magnetic field distribution around a nanomagnetic element is approximated for a point-like magnetic moment (**a**), a one-dimensional magnetic needle (**b**) and a two-dimensional stadium-shaped object (**c**). These forms are simplified representations for the experimentally used nanomagnets of a typical size of $450 \times 150 \times 15\,\text{nm}^2$, indicated by the black line. The $5 \times 5\,\text{nm}^2$-sized cells for the calculation are highlighted (and magnified in **a** and **b**) as blue dots

In Fig. 3.12 an overview of the three simplifications mentioned above is displayed. The field around the magnet shows a dipole-like far field for all cases. However, details such as the distribution of 'magnetic charges' at the poles and the field strengths in close proximity to a magnet's edge make it obvious that for densely packed arrays, the size and shape of a nanomagnetic element have considerable impact on the magnetostatic coupling and the macrospin approximation might not be appropriate.

For accessing the net magnetic field exerted by the surrounding magnets that is acting on a selected nanomagnet within the lattice, for instance during local switching, see Sect. 6.2.1, an extraction of the local magnetic-field strengths is required. For doing so, an area is considered that corresponds to the position of the considered nanomagnet at which knowledge about the average magnetic-field strength is desired. The magnetic flux passing through that area is averaged to approximate the effective magnetic field that acts on that particular nanomagnet. Whether or not the considered magnet would switch in response to that effective field depends on the projection of the calculated net field onto the easy axis of the nanomagnet.

3.6 Uncovering Long-Range Order from Micromagnetic Images

As described in the introduction to MFM, see Sect. 3.2.1, the MFM signal from an array of in-plane magnetised Ising- or vortex-type building blocks arises from the out-of-plane magnetic stray-field gradients caused by the divergence of the magnetisation at surfaces, see the surface magnetic charges in Fig. 3.5b. The particular distribution

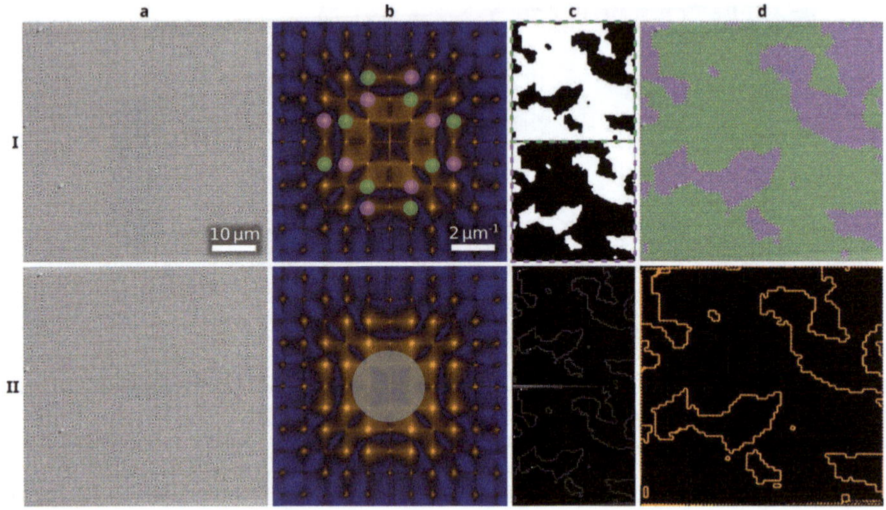

Fig. 3.13 Algorithmic recognition of toroidal order in nanomagnetic arrays. Protocols for the identification of domain- (**I**) and domain-wall states (**II**). **a** Sufficiently large MFM image, here with 70×70 unit cells. **b** A two-dimensional Fourier transform and inverse transforms are performed on selected regions (see coloured indications) that capture signatures of the $-T$ (violet, **I**) and $+T$ (green, **I**) toroidal domain states and of the domain wall (grey, **II**), respectively. **c** The subsequent binning of grey values enhances the contrast, while a following convolution with a Gaussian and application of thresholding yields two binary images of each of the domain sates (**I**) and of domain walls (**II**), respectively. **e** These images can now be used as a colour-coded background layer to the MFM images to emphasise the particular aspect of the order

of out-of-plane stray fields allows for an assignment of the magnetic state of the considered nanomagnetic element. Because a manual assignment of micromagnetic states in an array is laborious (yet, sometimes inevitable), I aimed for an algorithm-based solution.

A digital post-processing protocol has been developed to extract the ferroic domain structure directly from MFM images, using, for example, IMAGEJ/FIJI and GIMP. The protocol has been applied for the automatic recognition of long-range order in my main system, the Ising-type toroidal square array, see Fig. 4.1a. The sequence to reveal domain states is shown in Fig. 3.13I and the protocol to emphasise domain walls is shown in Fig. 3.13II.

References

1. Lehmann J et al (2019) Microdisplays as a versatile tool for the optical simulation of crystal diffraction in the classroom. J Appl Crystallogr 52(2):457–462. https://doi.org/10.1107/S1600576719001948

2. J. I. Martin et al. Ordered magnetic nanostructures: fabrication and properties. J Magn Magn Mater (2003), p. 53
3. Fitzsimmons MR, Silva TJ, Crawford TM (2006) Surface oxidation of Permalloy thin films. Phys Rev B 73(1):014420. https://doi.org/10.1103/PhysRevB.73.014420
4. H. Hopster et al., eds. Magnetic Microscopy of Nanostructures. Springer Berlin Heidelberg, 2005. http://dx.doi.org/10.1007/b137837
5. U. Hartmann. Magnetic Force Microscopy (1999), p. 36
6. E. Meyer, H. J. Hug, and R. Bennewitz. Scanning Probe Microscopy. Springer Berlin Heidelberg, 2004. https://doi.org/10.1007/978-3-662-09801-1
7. Zhu X et al (2005) Diffracted magneto-optical Kerr effects of permalloy ring arrays. J Appl Phys 97(10):10J712
8. B. Bhushan and H. Fuchs. Applied scanning probe methods. Springer, 2009
9. Eaton P, West P (2010). Atomic Force Microscopy Oxford University Press. https://doi.org/10.1093/acprof:oso/9780199570454.001.0001
10. O. Kazakova et al. Frontiers of magnetic force microscopy. J Appl Phys 125 6 (2019), p. 060901. http://aip.scitation.org/doi/10.1063/1.5050712
11. Y. Martin and H. K. Wickramasinghe. Magnetic imaging by "force microscopy" with 1000 Å resolution. Appl Phys Lett 50 20 (1987), pp. 1455-1457. http://aip.scitation.org/doi/10.1063/1.97800
12. Saenz JJ et al (1987) Observation of magnetic forces by the atomic force microscope. J Appl Phys 62(10):4293–4295
13. R. A. Buckingham. The classical equation of state of gaseous helium, neon and argon. Proceedings of the Royal Society of London. Series A. Math Phys Sci 168 933 (1938), pp. 264-283. http://dx.doi.org/10.1098/rspa.1938.0173
14. Garcia R (2002) Dynamic atomic force microscopy methods. Surf Sci Rep 47(6):197–301. https://doi.org/10.1016/S0167-5729(02)00077-8
15. Whangbo M-H, Bar G, Brandsch R (1998) Description of phase imaging in tapping mode atomic force microscopy by harmonic approximation. Surf Sci 411(1):794–801. https://doi.org/10.1016/S0039-6028(98)00349-5
16. Hartmann U (1989) The point dipole approximation in magnetic force microscopy. Phys Lett A 137(9):475–478. https://doi.org/10.1016/0375-9601(89)90229-6
17. U. Hartmann. Fundamentals and Special Applications of Non-Contact Scanning Force Microscopy. In: Advances in Electronics and Electron Physics. Vol. 87. Elsevier, 1993, pp. 49-200
18. https://www.ntmdt-si.com/products/modular-afm/ntegra-ii. [Online content, accessed on 27.12.2019]
19. Kerr J (1877) On rotation of the plane of polarization by reflection from the pole of a magnet. The London, Edinburgh, and Dublin Philosophical Magazine and Journal of Science 3(19):321–343. https://doi.org/10.1080/14786447708639245
20. P. Weinberger. John Kerr and his effects found in 1877 and 1878. Phil Mag Lett 88 12 (2008), pp. 897-907. https://www.tandfonline.com/doi/abs/10.1080/09500830802526604
21. J. Zak et al. Fundamental magneto-optics. J Appl Phys 68 8 (1990), pp. 4203-4207. http://aip.scitation.org/doi/10.1063/1.346209
22. A. K. Zvezdin and V. A. Kotov. Modern magnetooptics and magnetooptical materials. Institute of Physics Pub, 1997
23. S. Sugano and N. Kojima. Magneto-optics. Springer, 2000
24. H. Kronmueller and S. Parkin, eds. Handbook of Magnetism and Advanced Magnetic Materials. 1st ed. John Wiley & Sons, Ltd, 2007. http://dx.doi.org/10.1002/9780470022184
25. A. Hubert and R. Schaefer. Magnetic domains: the analysis of magnetic microstructures. Springer, 2009
26. McCord J (2015) Progress in magnetic domain observation by advanced magneto-optical microscopy. J Phys D: Appl Phys 48(33):333001. https://doi.org/10.1088/0022-3727/48/33/333001

27. J. A. Arregi, P. Riego, and A. Berger. What is the longitudinal magneto-optical Kerr effect? J Phys D: Appl Phys 50 3 (2017), 03LT01. http://dx.doi.org/10.1088/1361-6463/aa4ea6
28. Hulme HR (1932) The Faraday Effect in Ferromagnetics. Proc R Soc A: Math Phys Eng Sci 135(826):237–257. https://doi.org/10.1098/rspa.1932.0032
29. Argyres PN (1955) Theory of the Faraday and Kerr Effects in Ferromagnetics. Phys Rev 97(2):334–345. https://doi.org/10.1103/PhysRev.97.334
30. Freiser M (1968) A survey of magnetooptic effects. IEEE Trans Magn 4(2):152–161. https://doi.org/10.1109/TMAG.1968.1066210
31. Ebert H (1996) Magneto-optical effects in transition metal systems. Rep Prog Phys 59(12):1665–1735
32. Traeger G, Wenzel L, Hubert A (1992) Computer experiments on the information depth and the figure of merit in magnetooptics. Phys Status Solidi (a) 131(1):201–227. https://doi.org/10.1002/pssa.2211310131
33. Bader SD (1991) Smoke. J Magn Magn Mater 100(1):440–454
34. Z. Q. Qiu and S. D. Bader. Surface magneto-optic Kerr effect (2000), p. 14
35. Krinchik GS, Artemjev VA (1968) Magneto-optic Properties of Nickel Iron and Cobalt. J Appl Phys 39(2):1276–1278
36. Buschow K, van Engen P, Jongebreur R (1983) Magneto-optical properties of metallic ferromagnetic materials. J Magn Magn Mater 38(1):1–22. https://doi.org/10.1016/0304-8853(83)90097-5
37. G. Di and S. Uchiyama. Temperature dependence of Kerr rotation for Ni film. J Appl Phys 75 8 (1994), pp. 4270-4272. http://aip.scitation.org/doi/10.1063/1.355968
38. P. Oppeneer. Chapter 3 Magneto-optical kerr spectra. Handbook of Magnetic Materials. Vol. 13. Elsevier, 2001, pp. 229-422
39. H. Ebert and G. Schuetz, eds. Spin-orbit-influenced spectroscopies of magnetic solids. 466. Springer, 1996
40. J. F. Dillon and J. P. Remeika. Diffraction of Light by Domain Structure in Ferromagnetic $CrBr_3$. J Appl Phys 34 3 (1963), pp. 637-640. http://aip.scitation.org/doi/10.1063/1.1729321
41. Boersch H, Lambeck M (1964) Zur Beugung des Lichtes an Magnetisierungsstrukturen. Zeitschrift fuer Physik 177(2):157–163. https://doi.org/10.1007/BF01375333
42. Telesnin RV et al (1972) Diffraction of light in magnetic stripe-structure. Phys Status Solidi (a) 12(1):303–306
43. B. Kuhlow and M. Lambeck. Light Diffraction By Magnetic Domains. Physica B+C 80 1 (1975), pp. 374-380. http://dx.doi.org/10.1016/0378-4363(75)90079-0
44. B. Kuhlow. Light Diffraction By Magnetic Domain Gratings And Its Applications. J Magn Magn Mater (1980), pp. 391-394
45. Geoffroy O et al (1993) TMOKE hysteresis loops in Bragg diffraction from 2D patterns. J Magn Magn Mater 121(1):516–519. https://doi.org/10.1016/0304-8853(93)91258-9
46. Souche Y, Schlenker M, Dos Santos A (1995) Non-specular magneto-optical Kerr effect. J Magn Magn Mater 140–144:2179–2180. https://doi.org/10.1016/0304-8853(94)00847-7
47. van Labeke D et al (1996) Diffraction of light by a corrugated magnetic grating: experimental results and calculation using a perturbation approximation to the Rayleigh method. Opt Commun 124(5):519–528. https://doi.org/10.1016/0030-4018(95)00598-6
48. Suzuki Y et al (1997) Simple model for the magneto-optical Kerr diffraction of a reg ular array of magnetic dots. J Magn Magn Mater 165(1):516–519. https://doi.org/10.1016/S0304-8853(96)00605-1
49. Vial A, Van Labeke D (1998) Diffraction hysteresis loop modelisation in transverse magneto-optical Kerr effect. Opt Commun 153(1):125–133. https://doi.org/10.1016/S0030-4018(98)00188-6
50. Vavassori P et al (1999) Magnetic information in the light diffracted by a negative dot array of Fe. Phys Rev B 59(9):6337–6343. https://doi.org/10.1103/PhysRevB.59.6337
51. Grimsditch M et al (2001) Brillouin scattering and diffracted magneto-optical Kerr effect from arrays of dots and antidots (invited). J Appl Phys 89(11):7096–7100

52. T. Schmitte et al. Magneto-optical Kerr effects of ferromagnetic Ni-gratings. J Appl Phys 87 9 (2000), pp. 5630-5632. http://aip.scitation.org/doi/10.1063/1.1359789
53. Schmitte T, Westerholt K, Zabel H (2002) Magneto-optical Kerr effect in the diffracted light of Fe gratings. J Appl Phys 92(8):4524–4530. https://doi.org/10.1016/j.spmi.2004.01.004
54. Schmitte T et al (2003) The Bragg-MOKE: magnetic domains in Fourier space. Superlattices Microstruct 34(1):127–136. https://doi.org/10.1016/j.spmi.2004.01.004
55. Vavassori P et al (2004) Magnetization reversal via single and double vortex states in submicron Permalloy ellipses. Phys Rev B 69(21):214404. https://doi.org/10.1103/PhysRevB.69.214404
56. Westphalen A et al (2006) Magnetization reversal of micropattern Fe bar array: Combination of vector and Bragg magneto-optical Kerr effect measurements. J Magn Magn Mater 302(1):181–189. https://doi.org/10.1016/j.jmmm.2005.09.005
57. Vavassori P et al (2007) Chirality and stability of vortex state in Permalloy triangular ring micromagnets. J Appl Phys 101(2):023902. https://doi.org/10.1103/PhysRevB.69.214404
58. T. Verduci et al. Fourier magnetic imaging. Appl Phys Lett 99 9 (2011), p. 092501. https://doi.org/10.1063/1.3630049
59. Grimsditch M, Vavassori P (2004) The diffracted magneto-optic Kerr effect: what does it tell you? J Phys: Condense Matter 16(9):275–294. https://doi.org/10.1088/0953-8984/16/9/R01
60. Westphalen A et al (2007) Invited article: Vector and Bragg Magneto-optical Kerr effect for the analysis of nanostructured magnetic arrays. Rev Sci Instrum 78(12):121301
61. Lee M-S et al (2008) Extended longitudinal vector and Bragg magneto-optic Kerr effect for the determination of the chirality distribution in magnetic vortices. J Appl Phys 103(9):093913. https://doi.org/10.1063/1.2919160
62. B. Aktas, L. Tagirov, and F. Mikailov, eds. Magnetic nanostructures. 94. Springer, 2007
63. Caloz C et al (2018) Electromagnetic Nonreciprocity. Phys Rev Appl 10(4):047001. https://doi.org/10.1103/PhysRevApplied.10.047001
64. Baranova NB, Bogdanov YV, Zel'dovich BY (1977) New electro-optical and magneto-optical effects in liquids. Sov Phys Usp 20(10):870–877. https://doi.org/10.1070/PU1977v020n10ABEH005470
65. Ross HJ, Sherborne BS, Stedman GE (1989) Selection rules for optical activity and linear birefringence bilinear in electric and magnetic fields. J Phys B: At, Mole, Opt Phys 22(3):459–473. https://doi.org/10.1088/0953-4075/22/3/011
66. Rikken GLJA, Strohm C, Wyder P (2002) Observation of Magnetoelectric Directional Anisotropy. Phys Rev Lett 89(13):133005. https://doi.org/10.1103/PhysRevLett.89.133005
67. T. Arima. Magneto-electric optics in non-centrosymmetric ferromagnets. J Phys: Condense Matter 20 43 (2008), p. 434211. https://doi.org/10.1088/0953-8984/20/43/434211
68. Szaller D, Bordacs S, Kezsmarki I (2013) Symmetry conditions for nonreciprocal light propagation in magnetic crystals. Phys Rev B 87(1):014421. https://doi.org/10.1103/PhysRevB.87.014421
69. Tomita S et al (2018) Metamaterials with magnetism and chirality. J Phys D: Appl Phys 51(8):083001. https://doi.org/10.1088/1361-6463/aa9ecb
70. Cheong S-W et al (2018) Broken symmetries, non-reciprocity, and multiferroicity. Npj Quantum Materials 3(1):19. https://doi.org/10.1038/nmat1804
71. Nisoli C (2012) On thermalization of magnetic nano-arrays at fabrication. New J Phys 14(3):035017. https://doi.org/10.1103/RevModPhys.85.1473
72. X. Zhang et al. Understanding thermal annealing of artificial spin ice. APL Mater 7 11 (2019), p. 111112. https://doi.org/10.1063/1.5126713
73. Ke X et al (2008) Energy Minimization and ac Demagnetization in a Nanomagnet Array. Phys Rev Lett 101(3):037205. https://doi.org/10.1103/PhysRevLett.101.037205
74. R. F. Wang et al. Demagnetization protocols for frustrated interacting nanomagnet arrays. J Appl Phys 101 9 (2007), 09J104. https://doi.org/10.1063/1.2712528
75. Rougemaille N, Canals B (2019) Cooperative magnetic phenomena in artificial spin systems: spin liquids, Coulomb phase and fragmentation of magnetism - a colloquium. Eur Phys J B 92(3):62. https://doi.org/10.1140/epjb/e2018-90346-7

76. Morgan JP et al (2011) Thermal ground-state ordering and elementary excitations in artificial magnetic square ice. Nat Phys 7(1):75–79
77. Tannous C, Gieraltowski J (2008) The Stoner-Wohlfarth model of ferromagnetism. Eur J Phys 29(3):475–487. https://doi.org/10.1088/0143-0807/29/3/008

Chapter 4
Tailoring of the Sample System

For modelling and implementing toroidal order in artificial spin systems on mesoscopic length scales, a number of requirements have to be fulfilled. First, the distribution of magnetic moments within a single unit cell has to favour a compensated, vortex-like configuration. Second, this local magnetic arrangement has to couple to neighbouring unit cells to allow for a collective behaviour. This chapter introduces my model systems and explains the challenges in finding suitable parameters for their successful experimental realisation.

4.1 Selection of Artificial Spin Arrangements

Symmetry defines the magnetic point groups that allow for the emergence of a toroidisation in crystals, see Sect. 2.1.5 on page 17. The chosen arrangement of macrospins or magnetic vortices, as introduced in Sects. 2.2.2 and 2.2.3 from page 30 onwards, has to comply with those symmetry groups. In addition, magnetostatic interactions have to couple the toroidal moments of single unit cells in order to drive a ferroic phase transition that results in a uniform magnetic vortex state constituting the system's ground state.

All tailored arrangements of magnetic building blocks are magnetically compensated by design, so that a decomposition into an uncompensated (magnetic) and a compensated (toroidal) part of the spin arrangements is not required [1]. The ferrotoroidic model systems can be sorted into two classes: macrospin-based toroidal arrays and magnetic-vortex-based toroidal arrays, see Fig. 3.1. Since macrospins intrinsically exhibit large dipole-like magnetic stray fields, their strong coupling has already been utilised in a variety of mostly frustrated magnetic arrays, see Sect. 2.2.4. In contrast, magnetic vortices in nanomagnetic elements are magnetically compensated and, as a consequence, display much weaker stray fields that correspond to higher-order multipoles. Therefore, careful geometric arrangements are requested for achieving a magnetostatic coupling that allows for a uniform toroidisation in the array.

J. Lehmann, *Toroidal Order in Magnetic Metamaterials*, Springer Theses,
https://doi.org/10.1007/978-3-030-85495-9_4

In the macrospin arrays, the vortex structures have been realised by a segmented circular arrangement of mesoscopic Ising spins, see Fig. 3.1a. In a two-dimensional crystal, corresponding spin structures have to comply with the translational periodic lattice and are, thus, restricted to three-fold, four-fold and six-fold symmetries, see Fig. 4.1a–c. The magnetic-dipole coupling can be seen as a combination of an intra-vortex coupling for stabilising the local toroidal moment and an inter-vortex coupling that results in preferentially likewise-circling magnetic vortex states in neighbouring cells, as shown in Fig. 3.11.

In the magnetic-vortex arrays, each nanomagnetic element as such hosts a vortex-like spin configuration and, hence, a toroidal moment. It has been experimentally proven that the chosen geometry of sub-micrometre- to few-micrometre-sized equilateral triangles made from nickel [2, 3] or permalloy [4] can indeed host magnetic vortex states. Further, such a geometry provides magnetic stray fields that are large enough to remain detectable by MFM [2, 3, 5, 6]. Rounded edges at these equilateral triangles increase the stability range for the emergence of a magnetic vortex state [4]. In addition, an equilateral triangle constitutes a non-centrosymmetric building block. This can have experimental advantages regarding the switching behaviour in external magnetic fields, as it will be shown later. A suitable arrangement of the triangular-shaped elements, forming corner- or edge-connected networks, introduces a coupling that favours a uniform toroidal order, as shown in Fig. 4.1d–f.

The preference of the toroidal state can be further enhanced by cutting a hole into the centre of each triangle. By doing so, the magnetic-vortex state gets favoured in two aspects: Its internal energy gets reduced [8] by avoiding the presence of the energy-costly vortex core, see Sect. 2.2.3. Further, the topology of the building block becomes non-trivial by the introduction of the additional internal surface. For this constrained case, a single-domain state can not be achieved any more, since a directed magnetisation would require the spins to bend around the hole and induces two domain walls located at the poles, forming a so-called *onion state*. As a consequence, the magnetised state increases in energy and becomes less stable.

For the scope of my work, I restrict the discussion mainly to two representative nanomagnetic systems covering one macrospin-based (The toroidal square array, see Fig. 4.1a) and one vortex-based (The triangular system shown in Fig. 4.1d) array. The reason for this choice of model systems is the following: The toroidal square array, on the one hand, is based on a plain quadratic lattice and as such provides a magnetic dipole-dipole coupling energy between the building blocks that can be partitioned into two independent terms. This allows for a simple theoretical description that mimics two fundamental exchange-interaction terms in magnetic materials—a symmetric 180° and an antisymmetric 90° coupling. The triangular tiling, on the other hand, represents the class of arrays exhibiting inherent magnetic vortex states. Among the introduced variants, the chosen structure benefits from its structural inversion-symmetry breaking. The arrangement exhibits an advantageous character for collectively manipulating the magnetic configuration with external magnetic fields, as discussed in Section 6.3. Though, the toroidal square array constitutes the main research subject of this book, because its underlying couplings provide a greater flexibility for engineering the system and, furthermore, the strong MFM

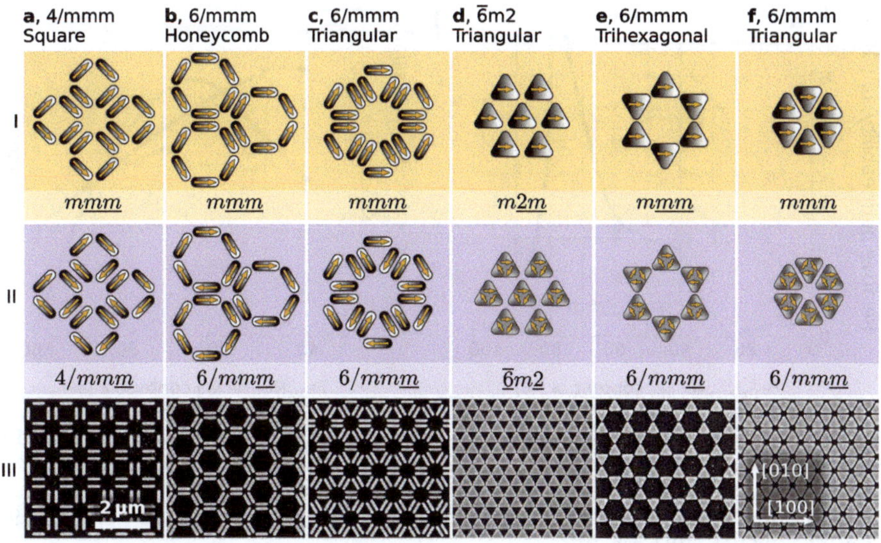

Fig. 4.1 Artificial spin systems with a magneto-toroidal order. Nanomagnet tilings and their corresponding structural point group which can host segmented (**a**–**c**) and inherent (**d**–**f**) magneto-toroidal moments. The internal magnetic state and the associated magnetic stray fields are indicated by gradients within nanoelements and by orange arrows. **I**, Sketch of metastable magnetised states and the corresponding magnetic point group [7] after the application of a horizontally aligned homogeneous magnetic field pointing to the right-hand side. **II**, Sketch of toroidal states and their corresponding magnetic point group with the toroidal moment pointing into the plane ($-t$). **III**, Electron-microscopy pictures of representing magnetic structures made from permalloy on silicon. Typical array sizes are of the order of $50 \times 50\,\mu m^2$. The coordinate system refers to the lattice orientations in line **III**

signal provided by macrospins simplifies the assignment of magnetic configurations. Note, however, that the results gained by the study of the two main systems can be generalised to other variants of nanomagnetic toroidal arrays as shown in Fig. 4.1.

4.2 Preparation of Building Blocks and Arrays

To study toroidal order at mesoscopic length scales, building blocks and lattice geometries have to be tailored to match a number of fundamental and experimental criteria. The optimisation regarding some of these criteria has been done as follows:

Array Size

For investigating long-range order in artificial ferroic arrays and in particular to avoid edge- or finite-size effects, arrays of at least a few hundred unit cells are reasonable

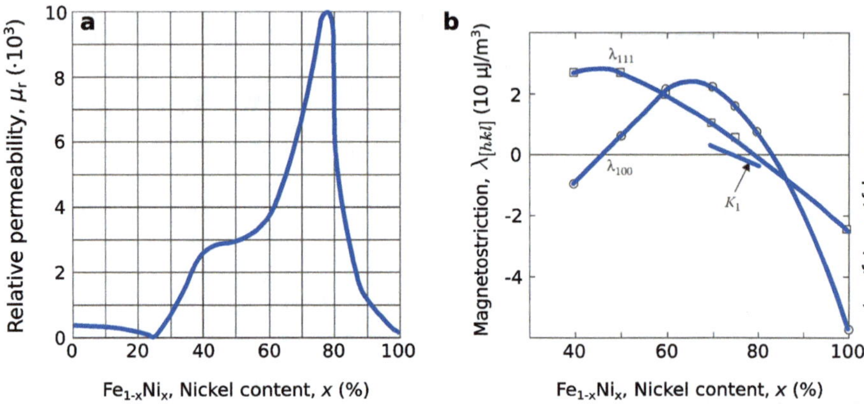

Fig. 4.2 Selection of magnetic properties of $Fe_{1-x}Ni_x$. **a** The relative permeability μ_r and **b** magnetostriction $\lambda_{[hkl]}$ and magnetocrystalline anisotropy $K^{(4)}$ of an iron-nickel alloy $Fe_{1-x}Ni_x$ as a function of x. For the composition around $Fe_{20}Ni_{80}$, the permeability shows a maximum, whereas the magnetostriction and magnetocrystalline anisotropy vanish. Figure adapted from Ref. [9], Chap. 12.2 and Ref. [10] Chap. 10.2

to be patterned and probed. For optical spectroscopy, arrays of size 3×3 mm have been fabricated, containing about 10^7 unit cells.

Growth Speed

The growth of nanomagnetic arrays mimics an annealing process and therefore has a strong impact on the formation of short- and long-range order, see Sect. 3.4. To get as-grown domains of a suitable size with respect to the array size, it is key to select an appropriate growth rate that suits multiple fabricated structures on a single substrate, see Sect. 3.1 on page 58. For most of my systems, a growth rate of ≈ 3 Å/s turned out to be apt.

Magnet and Substrate Materials

The magnetic material of choice is a soft magnetic nickel-iron alloy termed 'permalloy'. Close to its prototype composition of $Ni_{81}Fe_{19}$, it has outstanding magnetic properties such as a negligible magnetostriction, a large magnetic permeability and an extremely low magnetocrystalline anisotropy, see Fig. 4.2. The large permeability guarantees a strong response in a magnetic field, whereas the extremely low magnetocrystalline anisotropy and magnetostriction ensure a purely shape-determined magnetic state of microstructured permalloy elements, which is furthermore corroborated by the polycrystalline growth.

The substrates used in this study are (100)-oriented n-doped silicon chips of typically $8 \times 8\,\text{mm}^2$ lateral size and 200–500 μm thickness. These substrates are slightly conductive at room temperature and well-established for electron-beam lithography, and are highly transparent in the infrared spectral range. To allow for an optical analysis in the visible spectral region and for simultaneously reducing the refractive index of the substrate (which has a strong impact on the spectral position of plasmonic resonances), glass has been tested as an alternative substrate. Due to the highly insulating character of glass, however, charging effects have lead to difficulties in the lithography process and resulted in an unsuccessful fabrication process.

Size of the Building Blocks

The lateral dimensions of a nanomagnetic elements have to satisfy two main criteria: First, as already discussed in Sect. 2.2.1 on page 28, lower and upper size limits for Ising and for vortex elements of $\approx$10 nm thickness are given by the transitions to the superparamagnetic regime (below $\approx$50 nm) and to the multi-domain regime (above $\approx$5 μm), respectively. The micromagnetic measurement technique as such narrows these boundaries even further. Due to the minimum resolution of the used MFM measurement system of about 50 nm, the structural parameter should be slightly larger. Second, an envisaged practical use of the studied nanomagnetic arrays relies on linear optical phenomena and aims for effects in the visible or infrared region of the electromagnetic spectrum, see Sect. 2.3 on page 39. In order to be sensitive to an emergent toroidal state, the electromagnetic wave must have a wavelength λ of the order of the structural period a ($\lambda \approx a$, known as a *photonic crystal*) or may even be much larger than structural entities ($\lambda \gg a$, as the dipole approximation resulting in a so-called *effective medium*). In this context, the excitation of plasmonic resonances (more concretely surface- or particle-plasmon-polariton or lattice resonances) are known to be useful for enhancing light-matter interaction.

Thickness of the Building Blocks

A magnetic building block that is too thin leads to a small signal-to-noise ratio in MFM measurements due to weak stray fields. In addition, the magnetic state of a thin magnetic element becomes increasingly susceptible to perturbations induced by the moving magnetic tip during the MFM scan [11], see Fig. 4.3.

Distances Between the Building Blocks

For increasing the coupling between the nanomagnets, the patterning with high packing density is desirable. However, a merging of neighbouring elements has to be avoided. The minimal distance separating individual magnetic building blocks from each other depends on the quality and reliability of the lithography and lift-off

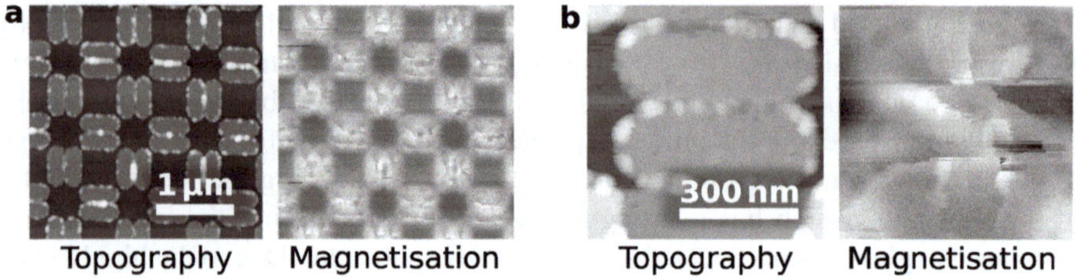

Fig. 4.3 Perturbed magnetisation distribution for too thin nanomagnetic elements. A 3-nm-thick permalloy structure of the toroidal square array (see Fig. 4.1a) has been imaged via MFM, line-by-line from top to bottom. Clearly visible is the perturbed magnetisation distribution as a consequence of the scan. **a** Section of the crystal lattice; **b** A pair of parallel neighbours in which sudden changes of the magnetic state are evident

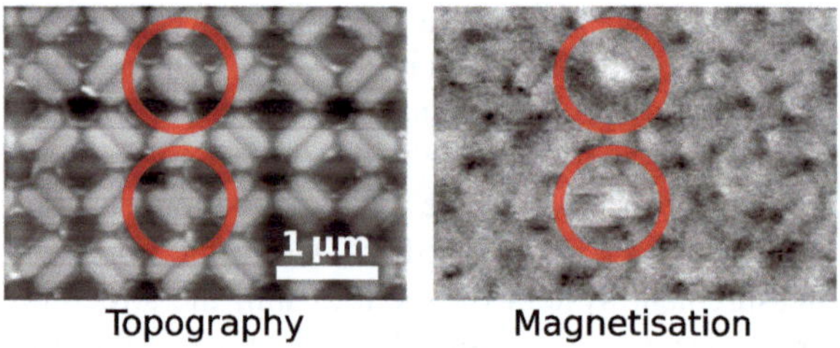

Fig. 4.4 Perturbed magnetisation because of exchange coupling. Local merging and, thus, distorted magnetic states (highlighted in red) occur for nanomagnets that are patterned with insufficient spacing between each other

process, see Sect. 3.1. For fabricating structures with precise dimensions, knowledge about synthesis-process parameters, such as the electron-beam focus, writing speed and the electron-beam current during lithography are crucial. Corresponding parameters have to be tested anew for every newly grown structure. For example, fluctuations of the electron-beam current result in fluctuations of geometric parameters and eventually to unwanted contact between the nanomagnets, see Fig. 4.4.

Aspect Ratio

To ensure that magnetic building blocks solely form their intended Ising or vortex state, a suitable aspect ratio is needed. For stable Ising states, an aspect ratio of $\geq$3:1 turned out to be practical for many cases. However, for excited states in the system (e.g. at domain walls), the actual magnetic configuration can deviate from the desired configuration, see Fig. 4.5. This is in particular significant for thick ($\gtrsim$20 nm) magnetic elements, for which the demagnetisation field yields a less stable Ising state, see Sect. 2.2.2.

Fig. 4.5 Non-Ising states in nanomagnetic arrays at a domain wall. Domain wall in an array made of ≥25-nm-thick magnets of 3:1 aspect ratio that locally display non-Ising behaviour at the position of a domain wall that separates $-T$ (violet) from $+T$ (green) domain states. The internal magnetisation of nanomagnets that form a domain wall falls into a flux-closed four-domain state instead of the expected single-domain state (for the corresponding transition, see Fig. 2.11)

4.3 Quantifying the Toroidisation in Nanomagnetic Arrays

The comparison of the toroidisation amplitude for a typical mesoscopic toroidal square array with two of the most-studied atomic candidate systems for ferro-toroidicity, $LiCoPO_4$ or $GaFeO_3$ gives noteworthy results. By calculating a magnetic moment of a single permalloy-based nanomagnet of size 450 nm × 150 nm × 15 nm to be about $m_{\text{mac}} \approx 8.7 \cdot 10^7\,\mu_B$, see Sect. 2.2.2, and assuming a typical geometry parameter $|\vec{r}| = 300\,\text{nm}$ (see Fig. 2.7 and Eq. 2.7) and a unit cell volume $V_{\text{uc}} = 1000 \times 1000 \times 15\,\text{nm}^3$, a calculation of the sample's toroidisation yields

$$|\vec{T}| = \left|\frac{1}{V_{\text{uc}}}\sum_i^4 (\vec{r}_i \times \vec{m}_{\text{mac},i})\right| \approx 700 \cdot 10^{-3}\,\mu_B/\text{Å}^2 \quad . \tag{4.1}$$

In contrast, the atomic systems provide values of $|\vec{T}|_{LiCoPO_4} = 6.2 \cdot 10^{-3}\,\mu_B/\text{Å}^2$ and $|\vec{T}|_{GaFeO_3} = 7.5 \cdot 10^{-3}\,\mu_B/\text{Å}^2$, respectively [1]. Hence, the toroidisation of the artificial nanomagnetic arrays is formally about a factor of 100 larger.

References

1. Ederer C, Spaldin NA (2007) Towards a microscopic theory of toroidal moments in bulk periodic crystals. Phys Rev B 76(21):214404. https://doi.org/10.1103/PhysRevB.76.214404
2. Yakata S et al (2010) Control of vortex chirality in regular polygonal nanomagnets using in-plane magnetic field. Appl Phys Lett 97(22):222503. http://aip.scitation.org/doi/10.1063/1.3521407

3. Jaafar M et al (2010) Control of the chirality and polarity of magnetic vortices in triangular nanodots. Phys Rev B 81(5):054439. https://doi.org/10.1088/0957-4484/19/28/285717
4. Vogel A et al (2012) Vortex dynamics in triangular-shaped confining potentials. J Appl Phys 112(6):063916. http://aip.scitation.org/doi/10.1063/1.4754418
5. Jaafar M et al (2008) Field induced vortex dynamics in magnetic Ni nanotriangles. Nanotechnology 19(28):285717. https://doi.org/10.1103/PhysRevB.81.054439
6. Krutyanskiy VL et al (2013) Second harmonic generation in magnetic nanoparticles with vortex magnetic state. Phys Rev B 88(9):094424. https://doi.org/10.1103/PhysRevB.88.094424
7. http://www.cryst.ehu.es/cryst/mpoint.html. (Online content, accessed on 29.12.2019)
8. Klaeui M et al (2003) Vortex formation in narrow ferromagnetic rings. J Phys: Condensed Matter 15(21):985–1024. https://doi.org/10.1088/0953-8984/15/21/201
9. Jiles D (2015) Introduction to magnetism and magnetic materials, 3rd edn. CRC Press. https://doi.org/10.1201/b18948
10. O'Handley RC (2000) Modern magnetic materials: principles and applications. Wiley
11. Kazakova O et al (2019) Frontiers of magnetic force microscopy. J Appl Phys 125(6):060901. http://aip.scitation.org/doi/10.1063/1.5050712

Chapter 5
Domains in Artificial Magneto-Toroidal Crystals

One of the two key requirements of n-dimensional ferroic systems is the spontaneous formation of homogeneously ordered areas, called domains, separated one from another by distinct $(n-1)$-dimensional entities, called domain walls, see Sect. 2.1.4 on page 16. These walls constitute natural interfaces within the system between two or more energetically degenerate realisations of a particular order. For the artificial nanomagnetic arrays investigated throughout this work, the spontaneous formation of long-range order connected to an order parameter is a hallmark for their classification as primary ferroic order [1]. Suitable symmetry groups allowing for toroidal order have already been identified a few decades ago and are listed in Sect. 2.1.6 on page 19. The nanomagnetic arrays that have been investigated here, see Fig. 4.1 on page 83, fall in one of these groups, setting the basis for their further scrutiny not only on the macroscopic but also on the microscopic scale.

Since ferrotoroidicity has been explored as a macroscopic phenomenon so far, microscopic origins of the ordering remain largely unexplored. In conventional materials, spins in a toroidally ordered crystal do not couple directly but via a fragile and well-balanced competition of different inter- and intra-exchange paths [2, 3]. These are typically variants of superexchange- or super-superexchange interaction via non-magnetic ligands, e.g., oxoanions (such as phosphates, silicates, germanates) [4]. For the present work, a substitution of microscopic exchange interactions with magnetic dipole-dipole interaction provides a direct geometric access to inherent coupling mechanisms, see for instance Fig. 2.13. Note that toroidal moments and toroidal crystals with a significant contribution from the magnetic-dipole interaction [5] have also been found and studied in systems with a ring-like arrangement of high-magnetic-moment rare-earth ions in molecular metal-organic frameworks [4], see Section 2.1.6. At this point, it is worth to recapitulate that ferroic order constitutes a macroscopic phenomenon without any constraints on the microscopic level. This allows for the engineering of a particular ferroic state using 'unconventional' constituents and arbitrary underlying interactions.

J. Lehmann, *Toroidal Order in Magnetic Metamaterials*, Springer Theses,
https://doi.org/10.1007/978-3-030-85495-9_5

This chapter presents the investigations on as-grown toroidal domain pattern in different artificial nanomagnetic crystals, see Fig. 4.1. The domain structures are surveyed experimentally via MFM. The main focus lies on the toroidal square array, see Fig. 4.1a. The chapter starts with a short study scrutinising the question of whether the long-range magnetic order that spontaneously forms in artificial magneto-toroidal arrays is pinned and pre-determined by structural inhomogeneities. This section is followed by a presentation of the measured spontaneous magnetic order in different types of macrospin-based nanomagnetic arrays. Hereafter, a statistical analysis of the local order is presented. The aim is to relate the observed domain structures with the energies and, thus, the probabilities of different excited states within the array. Doing so, the occupation density of domain-wall states as well as the respective domain sizes can be explained and—as a result—engineered at will by the initial design of the structure. Interestingly, the modification of the microscopic couplings not only influences the macroscopic properties such as the domain size, but also microscopic phenomena such as the density of emerging magnetic charges and the direction of magnetic moments that form the domain wall. These findings were supported with Monte-Carlo simulations by Amadé Bortis (see also Appendix C.5) and published most recently, see Ref. [6]. The chapter closes with a brief presentation of toroidal domain patterns in vortex-based nanomagnetic arrays.

5.1 The Influence of Structural Inhomogeneities

An initial step for the investigation of ferroic order is a test of the system's tendency for 'memorising' an as-grown domain structure due to structural defects, inhomogeneities or intrinsic strain profiles. In ferromagnets, for example, the pinning of magnetic order to localised defects has its origin in the minimisation of internal energy by correlating the position of domain walls with the position of defects such as impurities or grain boundaries via the magnetoelastic coupling. In the artificial nanomagnetic arrays investigated here, pinning based on magnetoelastic- or strain-induced effects can be excluded: The nanomagnets are well separated and not affected by a possible coupling mediated by strain via the substrate. However, any variation in the nanomagnet's thickness or lateral dimensions affects its blocking temperature, see Sect. 2.2.2. Therefore, spatial inhomogeneities, among others due to an incomplete lift-off during the fabrication (see Section 3.1) result in local hotspots (or 'coldspots') in the array (see, for example, Fig. 4.3). Those may serve as nucleation centres (individual macrospins of high blocking temperature), on the one hand, or leads to local nanomagnets with a smaller impact on the total energy of the system (individual macrospins of low blocking temperature) on the other hand.

The pinning and memory behaviour is investigated via a non-thermal relaxation protocol as introduced in Sect. 3.3 on page 70. Here, the thermal energy for 'activating' a magnetic system is substituted by the Zeeman energy induced from a spatially varying external magnetic field. In other words, the random and stochastic reorientation due to thermal energy is mimicked by a pseudo-random reorientation determined by

an alternating magnetic field. Thermal annealing and relaxation is now mimicked by a demagnetisation protocol to freeze a non-equilibrium multi-domain state that, in turn, can be compared with the as-grown multi-domain state.
In the experimental setup constructed for this, see Sect. B.1, the sample rotates with frequency $f_{\mathrm{rot}} \approx 167\,\mathrm{Hz}$ on a motorised spindle, oriented such that the sample plane lies in the plane of rotation. While rotating, the sample experiences a homogeneous magnetic field that oscillates with frequency $f_{\mathrm{osc}} \approx 50\,\mathrm{Hz}$ and is oriented in the sample plane as well. The oscillating magnetic field is generated by a stray-field transformer and can be varied in its flux density between zero and $\mu_0|\vec{\mathrm{H}}| \approx 120\,\mathrm{mT}$. The resulting field $\vec{\mathrm{H}}(t)$, is given by

$$\vec{\mathrm{H}}(t) = H_0 \left(1 - \frac{t}{t_{\mathrm{end}}}\right) \cdot \begin{pmatrix} \sin(2\pi f_{\mathrm{osc}}\, t)\, \sin(2\pi f_{\mathrm{rot}}\, t) \\ \sin(2\pi f_{\mathrm{osc}}\, t)\, \cos(2\pi f_{\mathrm{rot}}\, t) \\ 0 \end{pmatrix}, \qquad (5.1)$$

with H_0 as the initial field strength, t as the experiment time, and t_{end} as the total time for completing the field sweep. The magnetic-field amplitude in the sample plane is plotted in Fig. 5.1. As it can be seen, the magnetic field as a function of time provides a broad distribution of amplitudes and directions in the xy-plane. This allows for an 'activation' of differently oriented nanomagnets in the sample plane with a pseudo-random character and macroscopically imitates a thermal relaxation.

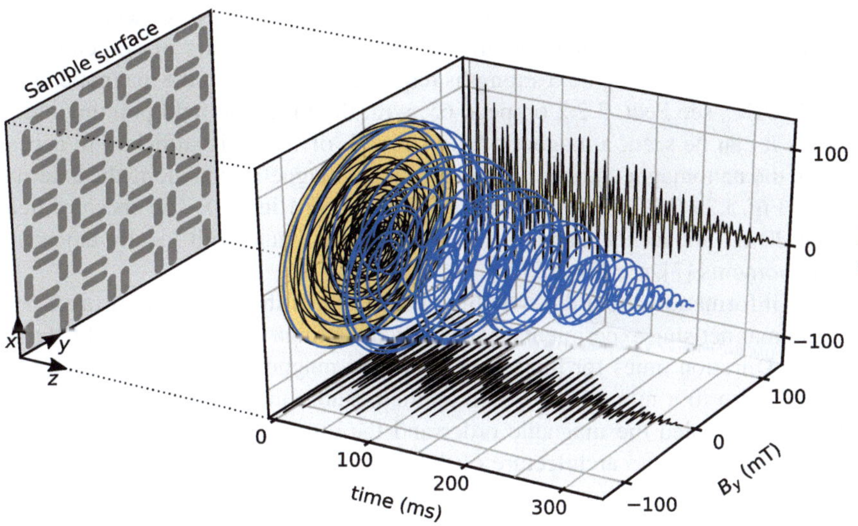

Fig. 5.1 Field distribution for a two-dimensional time-dependent demagnetisation. The sample rotates (frequency $f_{\mathrm{rot}} \approx 167\,\mathrm{Hz}$) in an oscillating magnetic field (frequency $f_{\mathrm{osc}} \approx 50\,\mathrm{Hz}$) with a flux density of up to $B \approx 120\,\mathrm{mT}$. The resulting field amplitudes in x- and y-direction (yellow-shaded curves) as well as the two-dimensional waveform (blue curve) with its projection onto the sample plane (yellow pattern) are shown for a total demagnetisation time of 330 ms

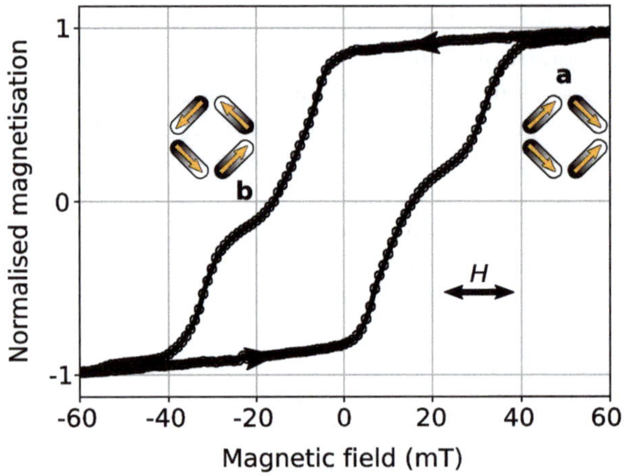

Fig. 5.2 Magnetisation curve for a toroidal square array. LMOKE measurements with an applied magnetic field along [110] reveal the switching behaviour of the nanomagnetic array. The remanent magnetised state **a** decays into a magnetically compensated toroidal state (step-like features in the magnetisation curve, **b** of random toroidal orientation by applying a magnetic field of 20–25 mT in the direction opposite to the net-magnetisation direction

I performed the study of domain pinning on the toroidal square array, see Fig. 4.1a. The array comprises $450\times150\,nm^2$ nanomagnets with a thickness of 20 nm and a lattice constant of 830 nm. The edge-to-edge separation distances of adjacent magnets are 60 nm for pairs of parallel neighbours and 110 nm for pairs of perpendicular neighbours. I measured the field strengths required for switching the magnetic order using LMOKE, see Sect. 3.2.2 on page 65, with the magnetic field pointing along [110]. As it can be seen, a magnetic field strength of 40 mT is sufficient to access and align the nanomagnet moments to induce a net magnetisation along the external field, see Fig. 5.2a (MOKE signal) and Fig. 5.3b (MFM image). The coercive field of about 20 mT induces a magnetically compensated state with locally emerging toroidal moments [7], see Fig. 5.2b.
Using the information about the switching behaviour of the considered array, I performed demagnetising experiments in a variety of settings with different field-sweep durations. Duration times for linear sweeps of the magnetic field in the range of a few seconds up to a minute have been applied. Subsequent MFM measurements, see Fig. 5.3, revealed the magnetic order and the alignment of the domain walls. The spatial position and architecture of the domain walls is found to be randomly distributed without indications for memory effects or pinning. Similar as in artificial spin-ice arrays [8, 9], the ground state (single-domain state) could not be reached easily. This can be explained by the small excitation energies of certain domain-wall states, which promotes their presence. Remarkably, the states obtained after application of the non-thermal relaxation protocols show no qualitative deviation from the as-grown state.

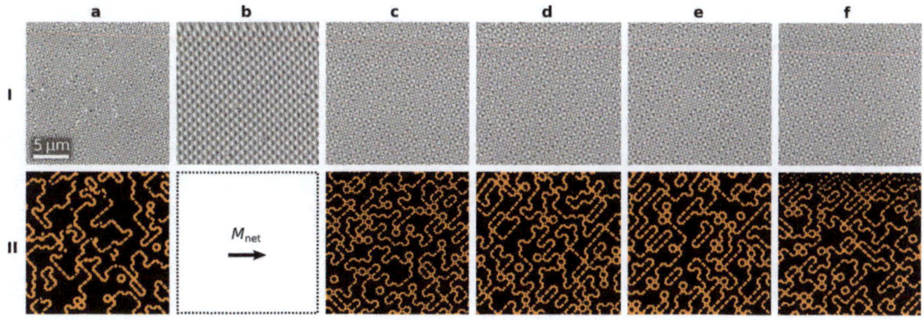

Fig. 5.3 Demagnetisation protocols applied to a toroidal square array. Experimentally obtained magnetic configuration (MFM, **I**), and extracted domain-wall configurations (**II**) are shown for the same area of the toroidal square array (see Fig. 3.13II). Displayed are the as-grown state **a** the remanent magnetised state **b** as well as four configurations obtained after demagnetising procedures **c**–**f**. The protocols differ in the duration of the field sweep (120 mT to zero): **c**, 1 min; **d** 10 s; **e** 3 s; **f** <1 s. The array shows no memory effect in its magnetic configurations

5.2 Revealing Domains in Ising-Type Nanomagnetic Arrays

In my arrays based on Ising-type nanomagnets, the magnetostatic interaction connects neighbouring magnetic moments in a way that a parallel orientation of toroidal moments constitutes the system's ground state, see Sect. 4.1 and Fig. 4.1a–c. Associated with this microscopic coupling is a collective behaviour that results in domains, obtained directly after growth. I confirmed the spontaneous formation of ferrotoroidic domains in three magneto-toroidal arrays of different geometry via MFM. Figure 5.4, shows the multi-domain patterns that exhibit areas of uniform toroidisation that extend over a large amount of unit cells. The successful visualisation of toroidal order at the mesoscale opened the door for further experiments.

5.3 Domain Engineering in Ising-Type Nanomagnetic Arrays

Control over the microscopic and macroscopic characteristics of domain patterns are key to the functional properties of ferroic materials. While the domain formation in magnetic materials with a net magnetisation is typically dominated by the minimisation of magnetostatic energy [10, 11] and has been the subject of many studies [11, 12], types of magnetic order with zero net magnetisation such as antiferromagnetism or ferrotoroidicity are gaining increasing attention. This is because those materials are considered to set the basis for an advanced memory and spintronic technology [13–20]. Due to their zero magnetostatic energy, the domain formation is no longer a consequence of the demagnetising field, so that less understood yet

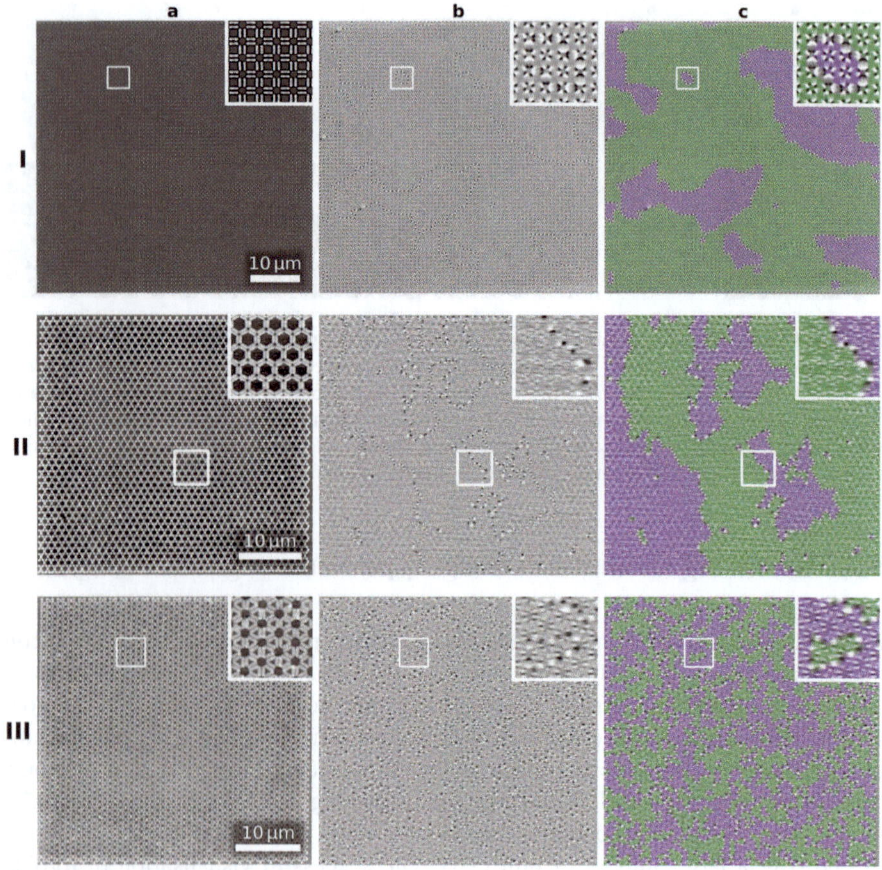

Fig. 5.4 Toroidal domain configurations in Ising-type nanomagnetic arrays. Three realisations of Ising-type nanomagnetic arrays (**I–III**) and their toroidal domain patterns, see also Fig. 4.1a–c. **a** The topography is imaged via AFM. **b** The magnetic configuration is measured by MFM. **c** Post-processing, see Sect. 3.6, reveals the toroidal domain patterns ($-\vec{T}$: violet, $+\vec{T}$: green) for each of the structures. The insets provide a close-up from certain areas

equally fundamental factors [21, 22] come to the fore—most importantly the role of microscopic interactions as intrinsic factors that promote collective behaviour in any magnetically ordered material.

As described above, the microscopic couplings that drive ferrotoroidicity in natural atomic crystals, such as $LiCoPO_4$, $LiFeSi_2O_6$ or $GaFeO_3$, are based on the fragile balance of competing energy terms. An adjustment and re-balancing of these energy terms provides novel insights into the peculiarities of the toroidal state. However, tuning of interaction energies in the above-mentioned materials, for example by epitaxial strain in thin films, is likely to result in major structural changes instead of the targeted modification of the fragile magnetic order. To my knowledge, tuning

of the interaction energies may be applicable only in the magnetic-dipole-coupled rare-earth molecular clusters [23], in which the balance of intra- and inter-toroidal coupling terms may be modified geometrically by modifying the ligand size.
Importantly, in my mesoscopic magneto-toroidal arrays, such a tuning is possible across a wide range without changing the overall symmetry or structure of the investigated system. For doing so, a mere modification of distances between macrospins within the array is required. The magnetic-dipole interaction energy $E_{\mathrm{D}}(r)$ that promotes the coupling between macrospins scales inverse-cubically with the distance r between the magnets. For the coupling of macrospins in the toroidal square array, both the interaction energy and the magnetic field strength can be separated in two independent terms, see Fig. 2.13 and Eq. 3.8. These two terms, elaborated above and in Sect. 3.4, constitute a classical analogue for a 90°- and a 180°-type of exchange interaction. These two interactions promote an intra- ($E_{\perp}$) and an inter- ($E_{\parallel}$) toroid coupling, respectively, see Fig. 3.11 on page 73.
Figure 5.5 shows the magnetic-field strength and interaction energy as a function of distance between magnetic moments (approximated as point dipoles) in the toroidal square array. Such a geometric tuning scheme is a very sensitive handle for controlling toroidal domain formation, since even small variations in the separation distance drastically influence the resulting coupling strength. In the following, I demonstrate the modification of microscopic as well as macroscopic properties of the toroidal square array by varying the local magnetic fields and corresponding coupling energies.
I compared the as-grown order in six magneto-toroidal arrays using MFM. These arrays are built from nanomagnets of identical size ($450 \times 150 \times 12\,\mathrm{nm}^3$), arranged with the same lattice constant of $1\,\mu$m. Yet, the arrays differ with respect to the nanomagnet positions within a unit cell, described by the geometry parameter r_{t}, introduced in Fig. 5.5. All arrays under comparison have been fabricated in a single growth process on the very same chip. This ensures a consistent quenching process for all arrays under consideration from the 'high-temperature' phase (macrospin thickness $\lesssim$3.5 nm) to the ordered phase (macrospin thickness $\gtrsim$3.5 nm) [24], see Sect. 3.4, and accordingly the formation of short- and long-range order with identical initial conditions.

5.3.1 Statistical Survey of Local Ordering

As introduced in Sect. 3.4 on page 72, the local order in the toroidal square array can be captured by considering vertices in the lattice at which eight nanomagnets meet, see Fig. 3.11. Such a vertex is formed by four parallel and four perpendicular pairs of macrospins. Thus, vertex energies are composed of multiples of the fundamental interaction energies $E_{\perp}$ and $E_{\parallel}$. The number of head-to-head or tail-to-tail alignments of perpendicular- and parallel-neighbouring nanomagnets ($n_{\perp}$ and $n_{\parallel}$, respectively) quantifies each vertex energy, which can be estimated via Eq. 3.9 on page 72. The possible values $n_{\perp}$ and $n_{\parallel}$ of a vertex are constrained by topological

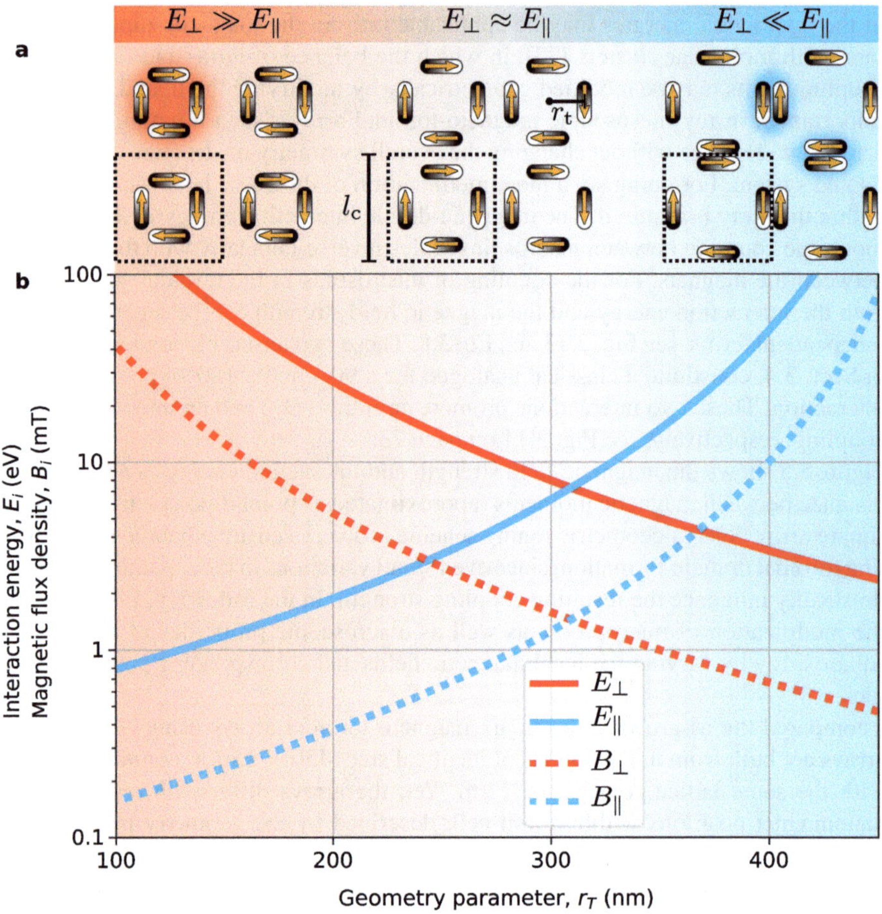

Fig. 5.5 Tuning microscopic interactions and magnetostatic coupling fields. a The magnetic coupling of neighbouring nanomagnets in the toroidal square array. Two different interaction terms $E_\perp$ and $E_\parallel$ can be tuned one against the other via the geometry parameter r_t. **b** The magnetic field that connects parallel pairs (blue) and perpendicular pairs (red) of macrospins and the corresponding interaction energy as a function of their respective separation. Here, a typical macrospin of magnetic moment $m = 8.7 \cdot 10^7\ \mu_B$ (As derived in Chap. 4 on page 81) and a fixed lattice constant $l_c = 1\ \mu m$ (see dashed squares) is assumed. Note that a variation of r_t by just 100 nm yields changes in the interaction energy and magnetic-field strength by as much as an order of magnitude

considerations. For instance, the total number of unfavoured configurations has to be an even number ($n_\perp + n_\parallel = \{0, 2, 4, 6, 8\}$). This formal consideration is equivalent with the obvious condition that a domain wall can not end abruptly somewhere (at a particular vertex) inside the array: an incoming domain wall at a certain point necessitates a second outgoing domain wall. Even though a total number of $2^8 = 256$ vertex states are distinguishable for the ensemble of eight nanomagnets, my nearest-

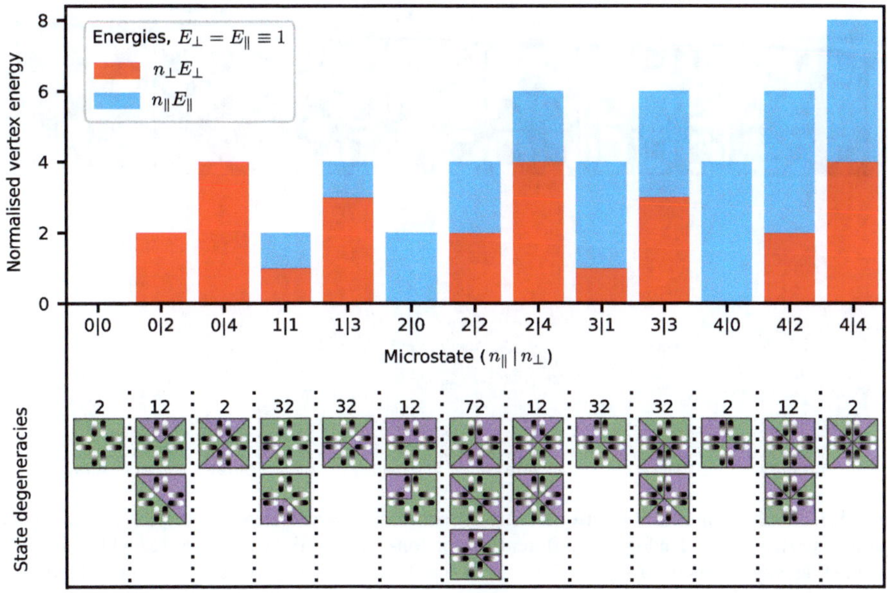

Fig. 5.6 Energy-state diagram for the macrospin-based toroidal square array. Thirteen different vertex states, parametrised by $(n_\parallel | n_\perp)$, of energy $E_{\text{vertex}} = n_\parallel E_\parallel + n_\perp E_\perp$ are distinguished, while the two interaction-energy terms E_i in the upper part of the diagram are set to 1. The lower part of the diagram quantifies the state degeneracy and displays a selection of magnetic configurations representing each state. The magnetisation direction of a macrospin is sketched as a black-to-white gradient, whereas the background colour relates to the toroidal moment in each domain state ($-\vec{t}$: violet, $+\vec{t}$: green), see Fig. 2.8

neighbour approximation results in thirteen degenerate vertex configurations that define all possible energy states in the system, see Fig. 5.6.

The statistical analysis that I present in the following calculates an energy for each of these thirteen vertex states based on the experimentally determined probability of their respective occurrence, see Eq. 3.10 on page 74. Therefore, an extracted set of vertex-state probabilities allows to relate back to the underlying balance between $E_\perp$ and $E_\parallel$. This estimation of effective pairwise interaction energies, related to the variation of r_t, can be used to sort the energy landscape of measured arrays into three distinct regimes ($E_\perp \gg E_\parallel$, $E_\perp \approx E_\parallel$ or $E_\perp \ll E_\parallel$, according to Fig. 5.5a). Displayed in Fig. 5.7a, are the six nanomagnetic arrays with different r_t, imaged pristine after growth. Each array has an area of about $50 \times 50\,\mu\text{m}^2$ and contains about 2500 unit cells. An area of $30 \times 30\,\mu\text{m}^2$ around the centre of each array[1] has been imaged by MFM, see Fig. 5.7b. I counted and categorised (see Table 5.1) all vertex states extracted from the MFM images, see Fig. 5.7c.

---

[1] Defects that are located close to the edges of some arrays prevented a comparative analysis of whole arrays.

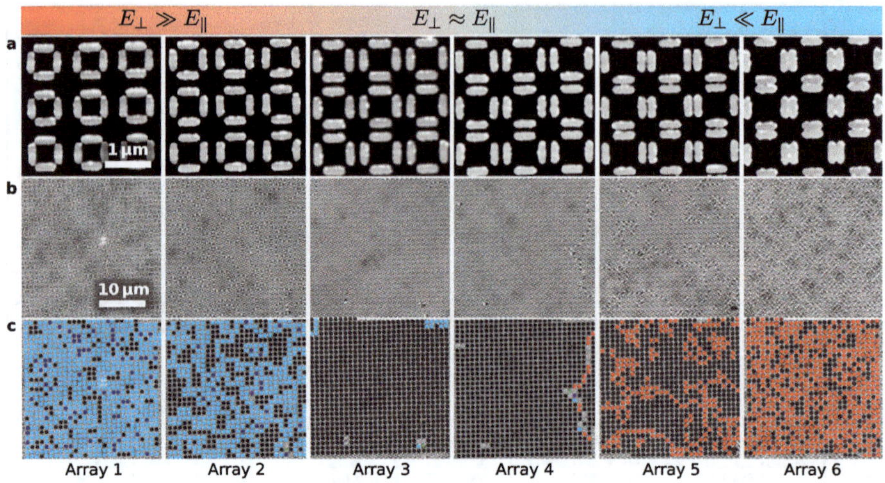

Fig. 5.7 Tuning microscopic interactions in toroidal square arrays. By varying the geometry parameter r_t, as defined in Fig. 5.5, different interaction-energy ratios can be realised with a constant magnet density across the arrays. **a** Sections of the AFM scans displaying 3 × 3 unit cells formed by four 450 × 150 × 12 nm^3 nanomagnets each. The apparent contact of adjacent magnets is due to an insufficient resolution of the AFM system. **b** MFM scans performed with a low-moment tip (see Tab. 3.1) showing approximately 30 × 30 unit cells. **c**, Assignment of vertex states according to Fig. 5.6 and Table 5.1

Table 5.1 Relation between measured vertex-state probabilities and calculated interaction energies. Vertex-state probabilities for each of the six measured macrospin-based toroidal square arrays. The occupation of each microstate, quantified by the number of head-to-head or tail-to-tail aligned neighbouring pairs of nanomagnets ($n_\parallel | n_\perp$) can be associated with vertex energies that depend only on $E_\perp$ and $E_\parallel$ and an assumed thermal energy E_{th} that corresponds to the growth temperature (room temperature, 26 meV). The vertex-state energy is calculated using Eq. 3.9, see Fig. 5.8. The statistical error for the number of counted vertex states N is approximated with $\sqrt{N}$. For simplicity, vertex states that do not occur in any of the arrays are not listed

$n_\parallel \mid n_\perp$		Array 1	Array 2	Array 3	Array 4	Array 5	Array 6
0|0 (%)		17.3 ± 1.5	43 ± 2.3	97.2 ± 3.5	96 ± 3.4	70.7 ± 3	31.5 ± 2
0|2 (%)		0	0	0	1.6 ± 0.4	28.8 ± 2	63.8 ± 2.8
0|4 (%)		0	0	0	0	0.25 ± 0.2	4.7 ± 0.8
1|1 (%)		0	0.4 ± 0.2	0.8 ± 0.3	1.95 ± 0.5	0.25 ± 0.2	0
2|0 (%)		75 ± 3	54.2 ± 2.6	1.9 ± 0.5	0.25 ± 0.2	0	0
2|2 (%)		0	0.1 ± 0.1	0.1 ± 0.1	0.1 ± 0.1	0	0
3|1 (%)		0	0.1 ± 0.1	0	0.1 ± 0.1	0	0
4|0 (%)		7.7 ± 1	2.2 ± 0.5	0	0	0	0
$E_\perp \mid E_\parallel$ (meV)		(205 | 5.5)	(167 | 20)	(117 | 74)	(79 | 97)	(36 | 185)	(13 | 210)

Following Eq. 2.3, the experimentally determined probability for a certain vertex state depends on the underlying state energy and the state degeneracy in relation to an estimated thermal energy. Therefore, the only free parameters are the two coupling energies $E_\perp$ and $E_\parallel$ and a thermal energy at which the system equilibrates. Since here, the growth of nanomagnetic arrays is performed at room temperature, a corresponding thermal energy of 26 meV has been assumed. In the beginning of the growth, magnetic stray fields and thus coupling energies in between nanomagnets are much weaker than in the final state, where they reach values in the 10^5 K range [9]. Therefore, the system undergoes a transition from the superparamagnetic regime into the so-called *blocked* regime, where the direction of magnetisation of each nanomagnetic element is frozen, see Sect. 3.4 on page 72. The interaction energies during this freezing can now be adjusted in order to reproduce the experimentally extracted occupation densities of vertex states $(n_\parallel|n_\perp)$, via Eq. 3.10, see Fig. 5.8. The approximated interaction energies are displayed in Fig. 5.9 together with the thermal energy at which the equilibrium configuration became frozen. Evident trends for the two energies can be observed that show a crossover in their magnitude between array 3 and array 4. This crossover and, consequently, the coverage of three distinct coupling regimes ($E_\perp \gg E_\parallel$, $E_\perp \approx E_\parallel$ or $E_\perp \ll E_\parallel$) is desired for further studies of the microscopic and macroscopic features that correspond to these regimes.

5.3.2 *Effect on the Long-Range Order: Domain Sizes*

Tuning microscopic interactions in a ferroic system not only changes the occurrence of different microstates as discussed above but also influences macroscopic characteristics such as the average domain size. The relation of the domain size with its underlying microscopic basis provides valuable knowledge for an according tunability. Yet, no publications exist that discusses domain-size tuning in toroidally ordered matter to the best my knowledge. One reason for this is the complexity of spin-spin interactions that stabilise ferrotoroidicity. Another reason is the difficulty to disentangle the variety of intrinsic and extrinsic (for instance internal strain profiles or defect-, segregation- or vacancy-density inhomogeneities that may modify the order due to electronic-state shifts or via the magnetoelastic effect) contributions that influence parameters such as the domain size. The significant advantage of my mesoscopic ferroic systems is the emergence of a macroscopic ferroic order parameter whereas the microscopic source of order is fully resolvable and—as already shown—can be tailored on demand.

In the previous section, I demonstrated the transitions between different intrinsic coupling regimes ($E_\perp \gg E_\parallel$, $E_\perp \approx E_\parallel$ or $E_\perp \ll E_\parallel$) throughout the series of toroidal square arrays with varying geometry parameter. These transitions do not alter the macroscopic symmetry of the system, which means it always retains its toroidally ordered ground state, yet the transition sensitively affect the resulting as-grown domain patterns. I extracted the domain patterns for the series of six toroidal square arrays using the algorithm as described in Sect. 3.6 on page 75. Extended toroidal

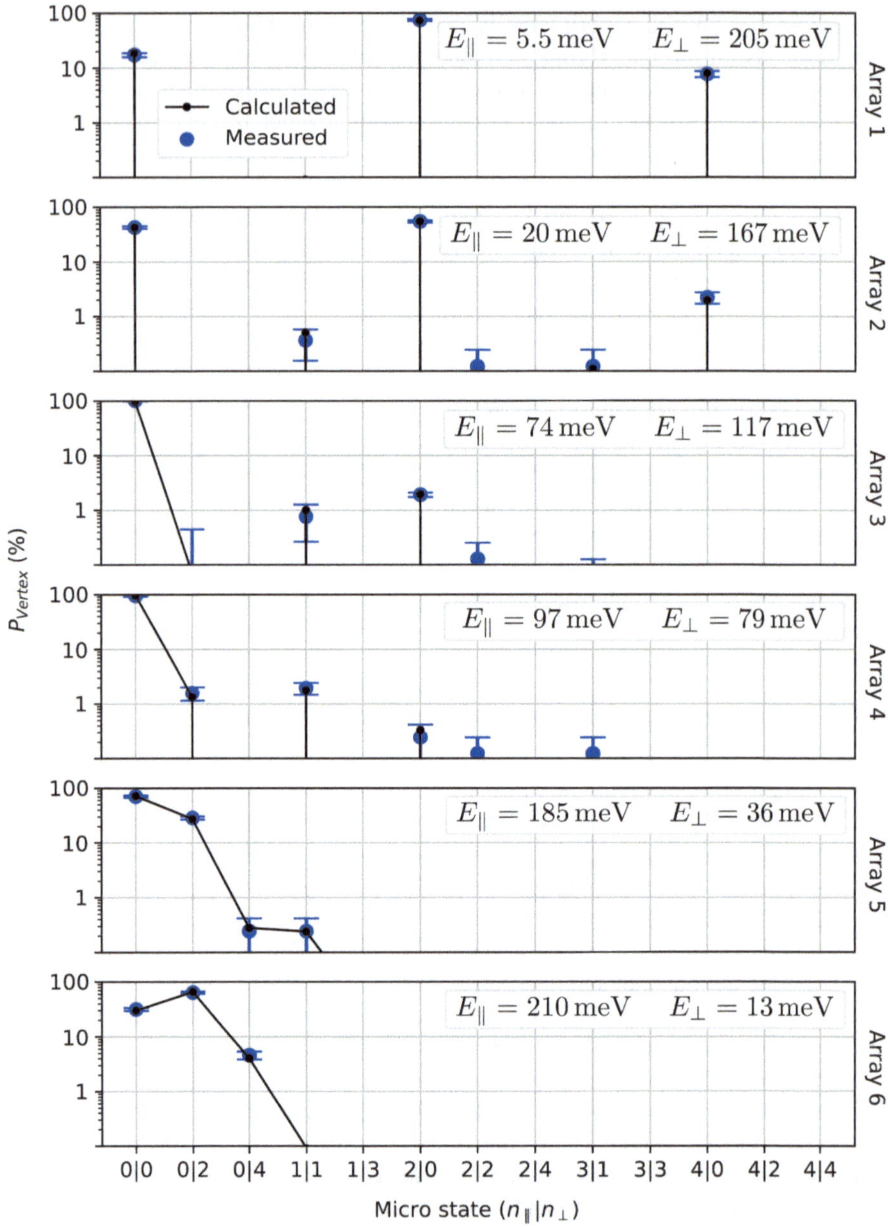

Fig. 5.8 Experimentally determined and calculated vertex-state probabilities in the series of toroidal square arrays. The data listed in Table 5.1 has been modelled via a Boltzmann-statistical approach using the two interaction-energy parameter $E_{\perp}$ and $E_{\|}$. Equation 3.10 has been applied to each vertex state that is quantified by the number of energetically unfavoured configurations of neighbouring nanomagnets ($n_{\|}|n_{\perp}$). A particular set of energy values allows to compute the expected probabilities (assuming a thermal energy of 26 meV) for all vertex states and, hence, to match the experimentally extracted probability. The statistical error for the number of counted vertex states N is approximated with $\sqrt{N}$

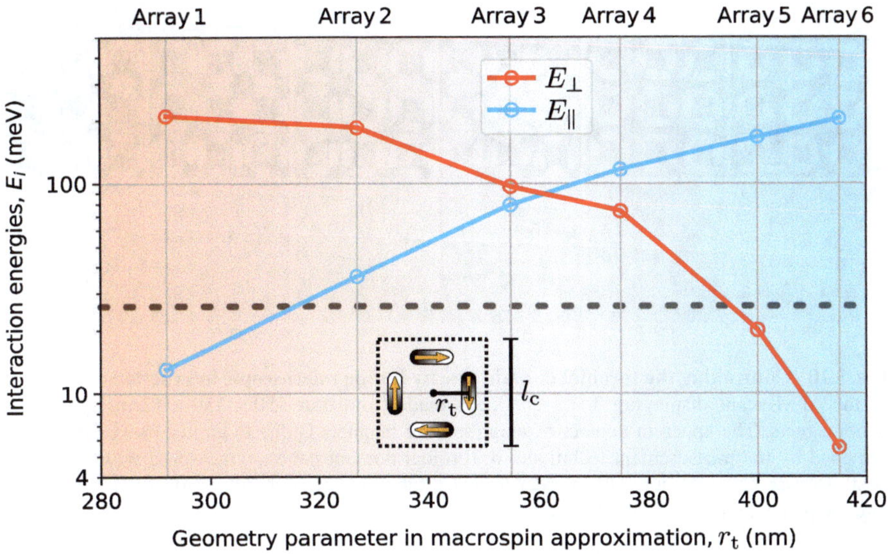

Fig. 5.9 Calculated interaction-energy parameters for six nanomagnetic arrays. The interaction energies have been extracted via a statistical analysis of the vertex-state probabilities, see Fig. 5.8. The geometry parameter r_t has been defined by virtually confining the nanomagnet's magnetic moment onto a point at its centre. The dashed grey line indicates the thermal energy (26 meV), corresponding to room temperature

domains can be observed, see Fig. 5.10. For large differences of the microscopic coupling energy (in particular array 1 representing $E_\perp \gg E_\parallel$ and array 6 representing $E_\perp \ll E_\parallel$), the smallest domain sizes emerge. The domain size increases when approaching equal interaction energies (array 3 and array 4 representing $E_\perp \approx E_\parallel$). The observed behaviour can be explained on the basis of the microscopic coupling energies extracted in the previous section. While the two-fold degenerate ground state ($\pm\vec{\mathrm{T}}$ as green and violet regions in Fig. 5.10b) can be seen as a state of zero internal energy, the excited states are separated from it by an energy that is determined by $E_\perp$ and $E_\parallel$. All those excited states in the system, see Fig. 5.6, correspond to different types of domain-wall states. Hence, their occupation density defines the proportion of domain-wall states with respect to domain states and thus relates to a particular average domain size. For the arrays at which one of the two interaction energies is relatively weak, the thermal energy enables a large occupation of the corresponding low-lying excited states. Hence, the comparably small domains in array 1 and array 6 are because of the small excitation energy of the first excited state (($n_\parallel|n_\perp$)=(0|2) and (2|0), respectively) and, furthermore, the state's high degeneracy (g=12).

A more generalised explanation can be made by considering the formation of stable magnetic subunits above the transition temperature T_c. One should remember that, during growth, the transition temperature is given by the growth temperature, whereas the interaction strength between magnetic building blocks increases due to

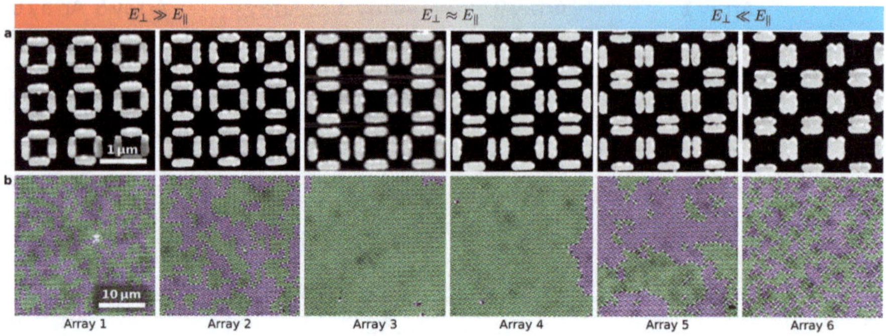

Fig. 5.10 Controlling the toroidal domain size by tuning microscopic interactions. a Sections of the AFM scans displaying 3×3 unit cells made from four $450 \times 150 \times 12\,\mathrm{nm}^3$ permalloy nanomagnets. The apparent contact of neighbouring magnets is due to an insufficient resolution provided by the measurement technique. **b** Toroidal domain patterns, measured with MFM and post-processed (see Section 3.6) in order to reveal the orientation of the toroidisation (violet: $-\vec{T}$, into the plane, green: $+\vec{T}$, out of the plane)

the increasing volume of magnetic material, see Sect. 3.4 on page 72. This motivates to use a more adequate term, such as the *transition thickness*. The subunits that form above T_c (see the arrangements highlighted in Fig. 5.5a) constitute squares of four macrospins in their flux-closed state connected via $E_\perp$ for $E_\perp \gg E_\parallel$ and pairs of antiparallel-aligned macrospins connected via $E_\parallel$ for $E_\perp \ll E_\parallel$, respectively. In both tuning regimes, the dominating interaction energy is larger than the thermal energy during the freezing of a 'high-temperature' equilibrium state. As a consequence, the arrangement of these subunits determines an emergent short-range order well above T_c. For characterising these subunits, a consideration of magnetic charges is helpful, which is often used for the description of artificial spin systems (see Sect. 2.2.4). Here, a dipole is formally reduced to a directionless entity formed by two opposite magnetic charges. Accordingly, the subunit arrangements consisting of four and two dipoles, represent eight and four alternating magnetic-charge configurations that mimic the magnetisation distribution of an octupole and a quadrupole, respectively. The interaction of octupoles for $E_\perp \gg E_\parallel$ or quadrupoles for $E_\perp \ll E_\parallel$ then promotes the phase transition to long-range order. As a consequence, the phase transition is driven by a fundamentally different microscopic source, which is of higher-order-multipole nature as compared with the retained dipole character for $E_\perp \approx E_\parallel$. Accordingly, the reduced coupling strength between the emergent magnetic subunits results in a shorter correlation length and, thus, smaller domains.

5.3.3 Effect on the Short-Range Order: Domain-Wall States

Domain walls are natural interfaces that form within a single-phase material. A variety of different realisations of domain walls has been observed in materials with a net magnetisation, see Fig. 2.5. Their microstructure is an important technological aspect and may relate to macroscopic measures such as the domain-wall mobility and inertia, see Ref. [11] Sect. 3.6.5 and therefore on the systems coercivity. However, gaining knowledge about domain-wall structures that occur in magnetically compensated long-range-ordered materials is a challenging task [25]. The experimental inaccessibility of the order in general and the nanoscale-small domain wall in particular hinders a comprehensive study of either the domain wall and an eventually present substructure within the wall. By using nanomagnetic arrays as an upscaled ferroic system, a close examination of the domain-wall morphology becomes possible. Furthermore, the relation between the interplay of local magnetic interactions and the realisation of a particular domain-wall provides new insights into the character of the ferrotoroidic order.
As introduced above, the tuning of microscopic interactions affects the occupation density of excited states (domain-wall states). It is therefore reasonable that the particular domain-wall type is another feature arising from the interplay of local energy terms. The left column of Fig. 5.11, with $E_{\perp} \gg E_{\parallel}$, reveals that the four nanomagnets of a unit cell exclusively form flux-closed magnetic octupoles. As a consequence, domain walls in this regime purely align along $\pm[100]$ or $\pm[010]$ and are located at the borders of the unit cell as defined in Fig. 5.5. Likewise, for $E_{\perp} \ll E_{\parallel}$, neighbouring parallel nanomagnets exclusively align anti-parallel and form magnetic quadrupoles. Hence, domain walls solely align through the defined unit cell along $\pm[110]$ and $\pm[\bar{1}10]$, see the right column of Fig. 5.11. For $E_{\perp} \approx E_{\parallel}$, both types of domain-wall states can be observed, see the centre column of Fig. 5.11.
In out-of-plane magnetised magnetic thin films with Bloch-type domain walls, see Fig. 2.5e–f, the magnetisation points parallel to the wall and thus perpendicular to the magnetisation direction within a domain. It may now be expected that toroidal domains consisting of unit-cell-sized magnetic whirls are encircled by domain walls that follows this magnetic-moment direction along the wall (referred to as a like-handed wall). Indeed, this scenario is observed for $E_{\perp} \gg E_{\parallel}$, where the walls align between two regions of connected four-nanomagnet octupoles exhibiting different toroidal moments and follow the magnetisation direction given by the edges of these subunits, as it is depicted in the left panel of Fig. 5.11d. However, this assumption about the sense of magnetic-moment circulation associated with the domain-wall in a toroidally ordered system is just a particular case that exists besides other possible realisations. It is neither true for systems with $E_{\perp} \approx E_{\parallel}$ nor for $E_{\perp} \ll E_{\parallel}$. Surprisingly, for $E_{\perp} \ll E_{\parallel}$, the wall is related to the opposite sense of circulation of magnetic moments around the wall compared with the sense of circulation inside the domain that is encircled by it (referred to as an opposite-handed wall), see the right panel of Fig. 5.11d. As a third possibility occurring in systems with $E_{\perp} \approx E_{\parallel}$, a combination of both wall types has ben observed. In this case, the magnetic moments

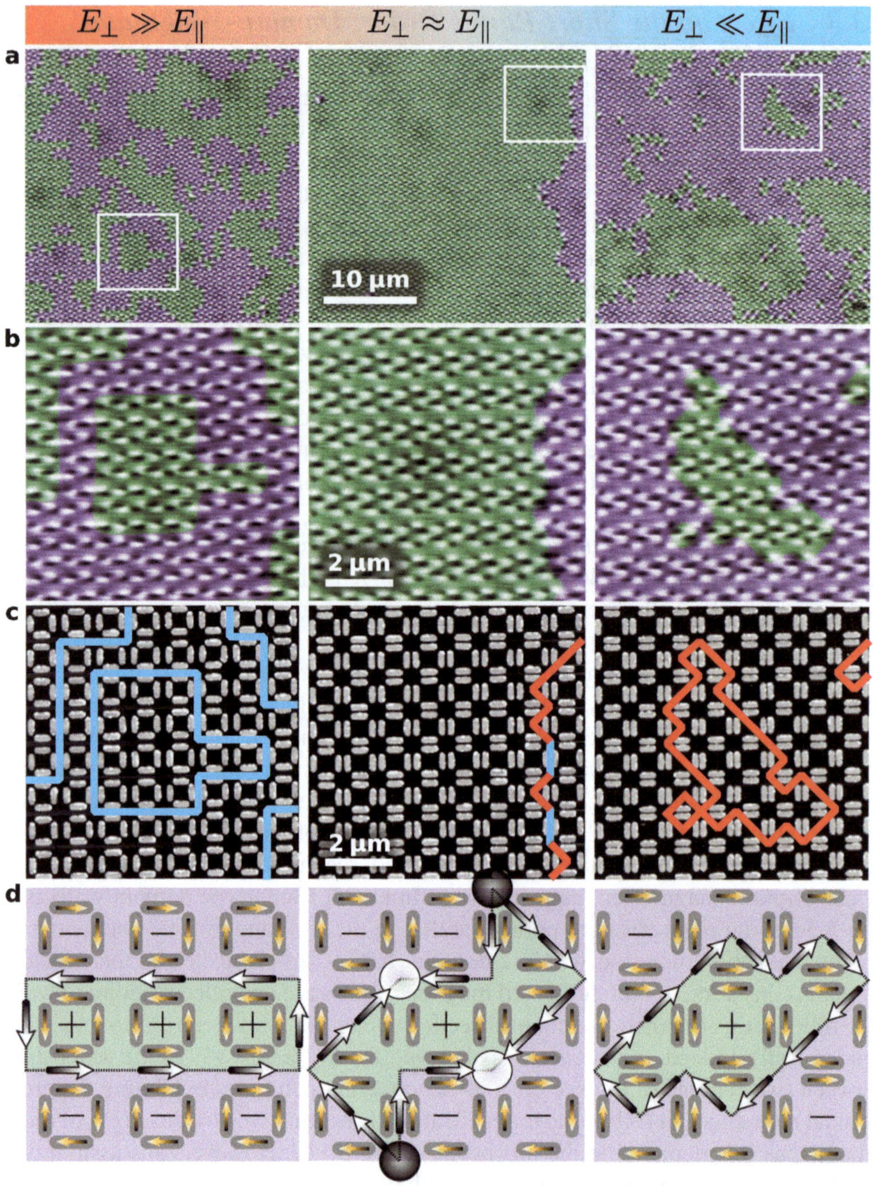

Fig. 5.11 Competing microscopic couplings affecting the domain-wall morphology. a Domain patterns revealed by post-processed MFM scans (see arrays 2, 4 and 5 from Fig. 5.10). Displayed are 30×30 unit cells each. The colour relates to the orientation of the toroidisation (violet: $-\vec{T}$, green: $+\vec{T}$. **b** Highlighted sections of 8×8 unit cells around a $+\vec{T}$ domain. **c** Topography of the emphasized sections, revealed by AFM. Sketched is the alignment of the domain wall with a color that relates to the particular cost of energy (blue: $E_\perp$, red: $E_\parallel$). **d** The magnetisation direction at a domain-wall. For $E_\perp \gg E_\parallel$ (left) the counter-clockwise circulating magnetic moments within the domain (orange gradient arrows) are accompanied by a magnetic-moment circulation of the same sense at the domain wall (black-white gradient arrows). In contrast, for $E_\perp \ll E_\parallel$ (right), the sense of magnetic-moment circulation at the domain wall opposes that observed in a particular domain. For $E_\perp \approx E_\parallel$ (centre), the magnetic-moment circulation of the domain wall alternates and yields localised magnetic charges (black and white discs). The relation between these spatial characteristics and the array's magnetic coercivity is shown in Appendix C.4

that form the wall alternate between like- and opposite handed with respect to the magnetic-moment circulation inside the encircled domain.

Those states at the positions where the magnetisation direction related to the domain wall reverses exhibit an unbalanced number of inwards to outwards pointing magnetic moments per vertex (see for example $(n_{\parallel}|n_{\perp}) = (1|1)$ in Fig. 5.6). Accordingly, these states can be seen as zero-dimensional topological defects that carry a localised effective magnetic charge [26–29]. These magnetic charges show similarities to *Bloch points* and are constrained to appear always in even numbers for magnetic charge equalisation, see the centred panel of Fig. 5.11d. The particular sense of circulation of magnetic moments at the domain-wall section that connects two topological defects comprising of opposite magnetic charges is protected. Hence, only the annihilation or emergence of opposite magnetic charges allow for a local change in the magnetic-moment circulation at the domain wall.

With respect to the toroidisation as the system's order parameter, the domain walls constitute two different types of boundaries from one domain to the other. For $E_{\perp} \gg E_{\parallel}$, the order parameter changes abruptly from one to the other unit cell, corresponding to the domain wall as sketched in Fig. 2.5a on page 15. For $E_{\perp} \ll E_{\parallel}$, the domain wall crosses unit-cell-sized toroidal moments and therefore bisects the net value in close proximity. Accordingly, this particular domain-wall type corresponds to the type of boundary sketched in Fig. 2.5b. Moreover, in analogy to the domain-wall types that occur in ferromagnetic materials, the toroidal domain walls show similarities with Bloch-type walls for which two possible realisations exist (see Fig. 2.5e–f). For a ferromagnetic film with out-of-plane-oriented magnetic domains, such Bloch-type walls are characterised by a net magnetisation direction that points along the wall. While in ferromagnetic materials, effects such as the famous Dzyaloshinskii-Moriya interaction are required for splitting the degeneracy of the two domain-wall configurations,[2] here the particular realisation can be tuned and manipulated at will by the interplay of local coupling terms.

5.4 Domain Patterns in Vortex-Type Magneto-Toroidal Arrays

The microscopic interaction that promotes ferrotoroidicity and toroidal domain states in vortex-element-based mesoscopic arrays is—similar as in the macrospin—based arrays—the magnetostatic coupling between the fundamental building blocks. However, whereas in macrospin-based arrays, the single-domain building blocks provide magnetic stray fields that are to leading order magnetic-dipole in nature, stray fields from magnetic vortex states are of higher-order multipole character and generally much weaker. This inherently provides challenges in the design of a suitable geometry to achieve a collective behaviour between the constituents and therefore

[2] Remember that this asymmetric exchange interaction necessitates a non-centrosymmetric material, whereas the particular symmetry violation predetermines the sign of this energy spitting.

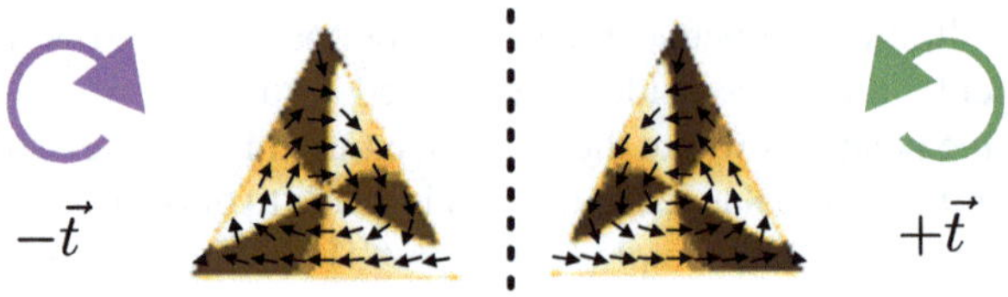

Fig. 5.12 Toroidal moment associated to the magnetic vortex state in a planar equilateral triangle. The spin configuration of a vortex state in micrometre-sized equilateral triangular structures represents a toroidal moment, see Sect. 2.1.5. Left: Clockwise whirled magnetisation distribution (arrows and gradients inside the triangle) corresponding to a toroidal moment of $-\vec{t}$. Right: Counter-clockwise whirled magnetisation distribution corresponding to a toroidal moment of $+\vec{t}$. In both configurations, measured with MFM, six magnetic poles are visible and located in pairs at each of the three corners of the triangle. Figure adapted from Ref. [30]

the stabilisation of ferroic order. As explained in Sect. 2.2.4 on page 34, surface magnetic charges are introduced by breaking the ∞-rotational symmetry of a circular disc that hosts a magnetic vortex state. The particular configuration of these charges allows for advanced inter-vortex coupling schemes. I decided for an equilateral triangle made of soft magnetic permalloy as a fundamental building block, as introduced in Sects. 3.1 and 4.1. This geometry, tailored to suitable dimensions, hosts a magnetic vortex state and exhibits magnetic fields that are of sufficient strength for an inter-element coupling and the analysis of the magnetic state via MFM. An example for the expected stray-field pattern gained with MFM is shown in Fig. 5.12. As introduced in Fig. 4.1d–f, different configurations of corner-connected and edge-connected triangles can be exploited. Each of these tilings allows for a magnetostatic coupling between the triangles that stabilises a ground state with a uniform sense of magnetic whirling. The array shown in Fig 4.1d has been utilised in most of the experiments since it has major advantages over the other array types regarding the manipulability of magnetic order and toroidal order as it will be shown in the following.

Arrays of nanomagnetic elements made of 20-nm-thick permalloy (passivated by 2 nm gold) and with corner radii of 50 nm have been investigated. Magnetisation curves, measured by LMOKE, see Sect. 3.2.2, have been used to verify the formation of magnetic vortex states within the structures, see Fig. 5.13. The averaged magnetisation curves reveal the minimum lateral dimensions for the triangles to retain the emergence of magnetic vortex states. Here, the array consisting of triangles with an edge length of 400 nm exhibits distinct plateaus within its hysteresis curve. These plateaus are well known indications for the emergence of a stable intermediate state during the magnetisation-reversal process in nanomagnetic elements. It is associated with the formation of a magnetic vortex state which forms abruptly, moves across the triangle and disappears jump-like at the other side. While in the array comprising of 300-nm-sized triangles this plateau is barely visible, it entirely disappears for the array made from 200-nm-sized triangles. Since the formation of toroidal domains in the triangle structures necessitates the existence of magnetic vortex states, only the 400-nm-sized building blocks (and possibly those of an even larger size) prove suitable as the constituents of the structure.

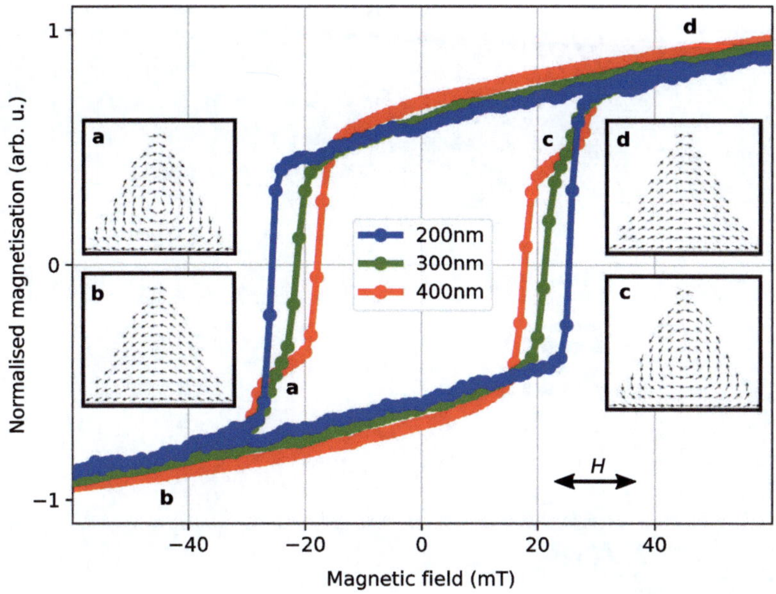

Fig. 5.13 Two-step magnetisation curves indicating magnetic vortex states. The magnetisation reversal has been measured by LMOKE (see Sect. 3.2.2) for three arrays (see Fig. 4.1d) consisting of equilateral 20-nm-thick permalloy triangles with edge lengths of 200 nm (blue), 300 nm (green) and 400 nm (red), and 50 nm corner radii and 100 nm separations between elements. A magnetic field *H* has been applied parallel to a triangle's edge (±[100] direction). Laser light of 405 nm wavelength has been focussed to a spot diameter of ≈30 μm to obtain an average LMOKE signal that stems from many triangles in the array. The steps or in the magnetisation curve (**a** and **c**) are a signature for the nucleation, propagation and annihilation of a magnetic vortex. The array made from 400-nm-sized triangles exhibits step-like features and, thus, indications for a magnetic vortex state. In contrast, the arrays made from 300-nm and 200-nm-sized triangles scarcely or do not show these signatures, respectively, but rather only single-domain states (**b** and **d**) and are thus not suitable for further studies of toroidal order

Toroidal domain structures have been examined by MFM, see Fig. 5.14. The toroidal-moment direction for each triangle has been decoded by a comparison with the stray-field distributions of clockwise and counter-clockwise whirled vortex states as shown in Fig. 5.12. The MFM scans revealed toroidal domain patterns in the two corner connected arrays under examination, see Fig. 5.14. The magnetic configuration of the array composed of edge-connected triangles, see Fig. 5.14III, shows a characteristic that deviates. While indeed areas with uniform toroidal moment occur, domain-wall-like entities form across the array at which the magnetic state of the triangles forms a single-domain state. This case demonstrates the close proximity of the array to a transition [31, 32] form the vortex state as the preferred state of the array to the single-domain state emerging as a excited state. This behaviour is not unexpected, since the Zeeman-energy contribution that results from the interaction of nanomagnets in densely packed arrays has been determined to favour the formation of an internal single-domain rather than a vortex state [33].

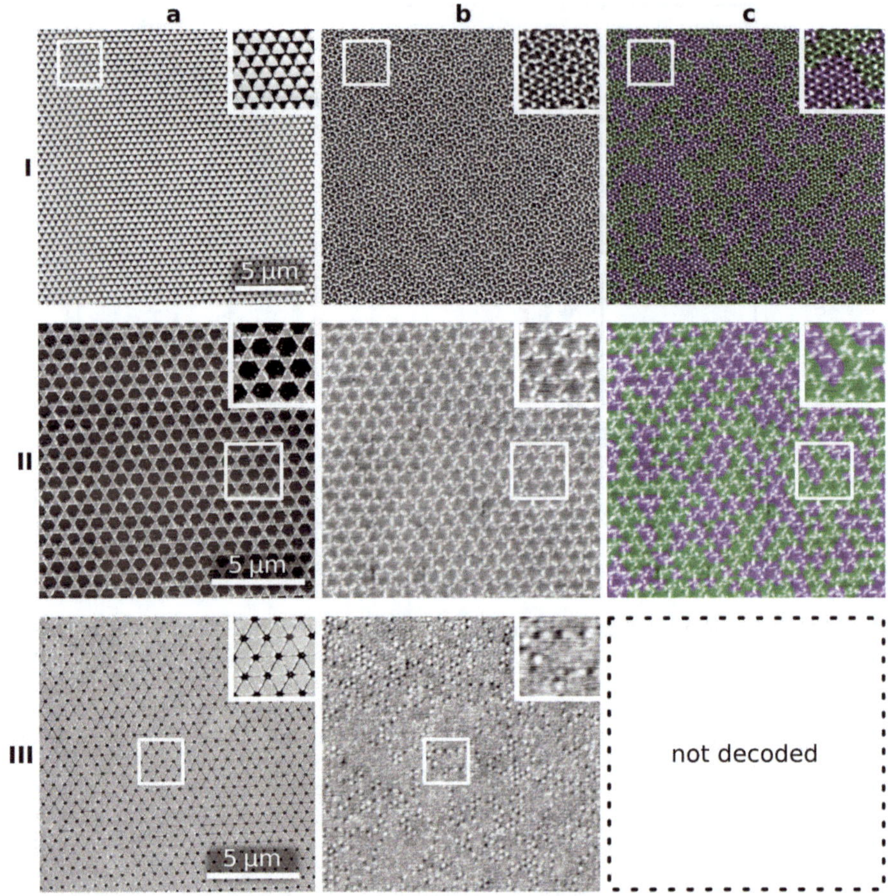

Fig. 5.14 Toroidal domain configurations in vortex-type nanomagnetic arrays. Three realisations of toroidal order based on triangular-shaped building blocks (**I–III**), see Fig. 4.1d–f. The edge lengths of the equilateral triangles are 400 nm with corner-curvature radii of 50 nm. The 20-nm- (**I**) and 15-nm-thick (**II–III**) elements are separated from each other by 100 nm (**I**) and 50 nm (**II–III**), respectively. **a** The topography is imaged via AFM. **b** The magnetic configuration is measured by MFM. **c** Post-processing reveals the toroidal domain configurations ($-\vec{T}$: violet, $+\vec{T}$: green). The order found in the array with the highest packing density (edge-connected triangles, **III**) indicates a proximity to a transition from a magnetic vortex to a single-domain state in each triangle. Hence, domain walls in this array comprise of non-compensated single-domain states within the triangles. The weak out-of-plane stray fields above the sample surface prevented the extraction of the underlying toroidal domain structure

5.5 Summary and Perspective

Long-range order in ferroic materials constitutes a main pillar of their functionality and is a key property in both fundamental research and device applications. While in the established primary ferroic classes the domain patterns are usually resolvable by a variety of techniques, see Ref. [34] chap. 4, in ferrotoroidic materials, the extrication of domains is challenging because of the lack of an experimentally established access to the order parameter. As a consequence, not much is known about the microscopic mechanisms that determine the formation and final configuration of toroidal domains. The upscaling from the atomic order to mesoscopic length scales opens up new possibilities to address these questions. With this work, I demonstrated ferrotoroidic multi-domain structures that spontaneously form in a set of different nanomagnetic arrays that are based on both magnetic single-domain and magnetic vortex-state base units. Initially, I made a clear distinction of the observed domains in toroidal metamaterials, which are of orientational nature, from the domains that have been found in artificial spin-ice arrays, which are of translational nature. The application of AC-demagnetisation protocols that mimic a thermal relaxation revealed a negligible memory effect of the domain patterns and no pinning of the magnetic order to structural inhomogeneities. A statistical analysis performed on the as-grown magnetic state of a series of arrays revealed the relation between microscopic (intra-toroidal and inter-toroidal) couplings and the probability for certain microstates that determine the macroscopic order. The experiments confirmed the existence of a transition between two coupling regimes that is accompanied by different types of short-range order. I found that the competition of microscopic couplings not only determines the particular path of the system towards order, but is in particular responsible for the resulting domain size and domain-wall morphology. This non-thermal handle for tuning the domain size is an interesting finding that is transferable to other systems with magnetic ordering mechanisms. Two types of domain walls have been identified whose appearance depends on the balance of the two microscopic interactions. These walls show similarities with Bloch-type domain walls in ferromagnetic systems. A detailed characterisation with respect to their alignment and their associated direction of magnetisation allowed the control of homo-'chiral' walls of either type. This can be seen as an analogue to the manipulation of the effective amplitude and sign of the Dzyaloshinskii-Moriya coupling vector in ferromagnets. A mixture of both wall types yields an alternating magnetisation direction inside the wall. This substructure is accompanied by the emergence of Bloch points as localised and topologically protected magnetic charges. In ferromagnets, those Bloch points are known to influence, e.g., temporal measures of the underlying magnetic texture such as the domain-wall mobility.

The tunability of fundamental parameters such as the domain size, the domain-wall type and the density of topological defects demonstrates not only the peculiarities of ferrotoroidic order as such but also the benefits that result from an upscaling of a physical phenomena to the more-accessible mesoscale.

Further work might address the finding that a ground state of the system could not easily be reached by demagnetisation experiments. The question is not answered, whether this is due to the non-zero (a)thermal population of excited states due to a demagnetisation that has been applied too fast for relaxing the system into its ground state or due to the presence of a 'detoroidisation' field that—similar as the demagnetisation field in ferromagnets—enforces a multi-domain state in thermal equilibrium at zero temperature. For those and further investigations, a quantification of the correlation length in toroidal domain patterns may be of importance. Evaluating for instance a pair-correlation function for parametrising the tendency of a ferroic model system to order in a ferro- (uniform) or an antiferro- (staggered) like configuration may be necessary. Here, the comparison of neighbouring toroidal moments, which can align parallel or anti-parallel would provide a suitable measure χ:

$$\chi = \frac{\#_{\text{bonds}}(\pm\vec{\mathrm{T}} \mid \pm\vec{\mathrm{T}})}{\#_{\text{bonds}}(\pm\vec{\mathrm{T}} \mid \mp\vec{\mathrm{T}})}, \tag{5.2}$$

where $\#_{\text{bonds}}$ accounts for the number of bonds between likewise-oriented $(\pm\vec{\mathrm{T}} \mid \pm\vec{\mathrm{T}})$ or oppositely oriented $(\pm\vec{\mathrm{T}} \mid \mp\vec{\mathrm{T}})$ pairs of toroidal moments. Consequently, 'anti-toroidal' order would yield $\chi \ll 1$, whereas toroidal order would yield $\chi \gg 1$. Another case of interest besides the discussion of ferrotoroidic order may be the investigation of vortex-type and macrospin-type arrays at their stability threshold. In particular, for both types of arrays I found a particular combination of size, aspect ratio and thickness that results in excited states (domain-wall states) with an unexpected behaviour. In these, the stray field provided by nanomagnets in close proximity to a domain wall tend to perturb and destabilise the magnetic configuration of the wall. The consequence is a spatially varying local Zeeman term in the energy configuration of nanomagnets forming the wall. This induces a local change in their spin configuration from Ising to vortex (or four-domain state) or vice versa, in order to lower the total energy of the system. Thus, as a consequence of the strong coupling in the array, the corresponding reduced or increased magnetic moment at a domain wall provides a delicate degree of freedom for a nanomagnetic system. This local phase coexistence is an effect that is absent in the common Ising-like approximations for nanomagnetic arrays and demands a more sophisticated treatment in, e.g., large-scale statistical simulations.

References

1. Wadhawan VK (2000) Introduction to Ferroic materials, 1st edn. CRC Press. http://dx.doi.org/10.1201/9781482283051
2. Spaldin NA, Fiebig M, Mostovoy M (2008) The toroidal moment in condensed-matter physics and its relation to the magnetoelectric effect. J Phys Condens Matter 20(43). https://doi.org/10.1103/PhysRevB.88.094429

3. Lee C et al (2014) Analysis of the difference between the pyroxenes $LiFeSi_2O_6$ and $LiFeGe_2O_6$ in their spin order, spin orientation, and Ferrotoroidal order. Chem Mater 26(4):1745–1750. https://doi.org/10.1021/cm5001413
4. Gnewuch S, Rodriguez EE (2019) The fourth ferroic order: current status on Ferrotoroidic materials. J Solid State Chem 271:175–190. https://doi.org/10.1016/j.jssc.2018.12.035
5. Harris AB (2010) A system exhibiting toroidal order. Phys Rev B 82(18). https://doi.org/10.1103/PhysRevB.82.184401
6. Lehmann et al J (2020) Relation between microscopic interactions and macroscopic properties in ferroics. Nat Nanotechnol 15(11):896–900. https://www.nature.com/articles/s41565-020-0763-9
7. Lehmann J et al (2019) Poling of an artificial magneto-toroidal crystal. Nat Nanotechnol 14(2):141–144. https://doi.org/10.1038/s41565-020-0763-9
8. Mengotti E et al (2008) Building blocks of an artificial Kagome spin ice: Photoemission electron microscopy of arrays of ferromagnetic islands. Phys Rev B 78(14). https://doi.org/10.1103/PhysRevB.78.144402
9. Nisoli C et al (2010) Effective temperature in an interacting vertex system: theory and experiment on artificial spin ice. Phys Rev Lett 105(4):047205
10. Kittel C (1949) Physical theory of ferromagnetic domains. Rev Modern Phys 21(4):541–583. https://doi.org/10.1103/RevModPhys.21.541
11. Hubert A, Schaefer R (2009) Magnetic domains: the analysis of magnetic microstructures. Springer
12. McCord J (2015) Progress in magnetic domain observation by advanced magneto-optical microscopy. J Phys D Appl Phys 48(33). https://doi.org/10.1088/0022-3727/48/33/333001
13. Sander D et al (2017) The 2017 magnetism roadmap. J Phys D Appl Phys 50(36). https://doi.org/10.1088/1361-6463/aa81a1
14. Gomonay EV, Loktev VM (2014) Spintronics of antiferromagnetic systems (Review Article). Low Temp Phys 40(1):17–35. http://aip.scitation.org/doi/10.1063/1.4862467
15. Wadley P et al (2016) Electrical switching of an antiferromagnet. Science 351(6273):587–590. https://doi.org/10.1126/science.aab1031
16. Jungwirth T et al (2016) Antiferromagnetic spintronics. Nat Nanotechnol 11(3):231–241. https://doi.org/10.1038/nnano.2016.18
17. Jungwirth T et al (2018) The multiple directions of antiferromagnetic spintronics. Nat Phys 14(3):200–203. https://doi.org/10.1038/s41567-018-0063-6
18. Baltz V et al (2018) Antiferromagnetic spintronics. Rev Mod Phys 90(1). https://doi.org/10.1103/RevModPhys.90.015005
19. Zelezny J et al (2014) Relativistic Néel-order fields induced by electrical current in antiferromagnets. Phys Rev Lett 113(15). https://doi.org/10.1103/PhysRevLett.113.157201
20. Zelezny J et al (2017) Spin-orbit torques in locally and globally noncentrosymmetric crystals: Antiferromagnets and ferromagnets. Phys Rev B 95(1). https://doi.org/10.1103/PhysRevB.95.014403
21. Gomonay HV, Loktev VM (2007) Shape-induced phenomena in finite-size antiferromagnets. Phys Rev B 75(17). https://doi.org/10.1103/PhysRevB.75.174439
22. Gomonay O, Kondovych S, Loktev V (2014) Shape-induced anisotropy in antiferromagnetic nanoparticles. J Magnetism Magnetic Mater 354:125–135. https://doi.org/10.1016/j.jmmm.2013.11.003
23. Hewitt IJ et al (2010) Coupling Dy3 triangles enhances their slow magnetic relaxation. Angewandte Chemie Int Edn 49(36):6352–6356. https://doi.org/10.1002/anie.201002691
24. Nisoli C (2012) On thermalization of magnetic nano-arrays at fabrication. New J Phys 14(3). https://doi.org/10.1088/1367-2630/14/3/035017
25. Khodenkov GE (2006) Bloch lines in domain walls of antiferromagnets. Tech Phys 51(7):949–951. https://doi.org/10.1134/S1063784206070243
26. Castelnovo C, Moessner R, Sondhi SL (2008) Magnetic monopoles in spin ice. Nature 451(7174):42–45. https://doi.org/10.1038/nature06433

27. Mol LA et al (2009) Magnetic monopole and string excitations in two-dimensional spin ice. J Appl Phys 106(6). https://doi.org/10.1063/1.3224870
28. Ladak S et al (2010) Direct observation of magnetic monopole defects in an artificial spin-ice system. Nat Phys 6(5):359–363. https://doi.org/10.1038/nphys1628
29. Loreto RP et al (2015) Emergence and mobility of monopoles in a unidirectional arrangement of magnetic nanoislands. Nanotechnology 26(29). https://doi.org/10.1088/0957-4484/26/29/295303
30. Yakata S et al (2010) Control of vortex chirality in regular polygonal nanomagnets using in-plane magnetic field. Appl Phys Lett 97(22):222503. http://aip.scitation.org/doi/10.1063/1.3521407
31. Cowburn RP et al (1999) Single-domain circular nanomagnets. Phys Rev Lett 83(5):1042–1045
32. Jubert P-O, Allenspach R (2004) Analytical approach to the single-domain-to-vortex transition in small magnetic disks. Phys Rev B 70(14). https://doi.org/10.1103/PhysRevB.70.144402
33. Guslienko KY et al (2001) Magnetization reversal due to vortex nucleation, displacement, and annihilation in submicron ferromagnetic dot arrays. Phys Rev B 65(2). https://doi.org/10.1063/1.1450816
34. Tagantsev AK, Cross LE, Fousek J (2010) Domains in Ferroic crystals and thin films. Springer, New York. http://dx.doi.org/10.1007/978-1-4419-1417-0

Chapter 6
Poling of Artificial Magneto-Toroidal Crystals

The basis for key applications of ferroic materials is the ability to reverse the orientation of the order parameter and, accompanied by that, the sign or direction of a physical vectorial or tensorial quantity [1, 2]. This characteristic allows one to distinguish otherwise symmetry-equivalent material classes such as pyroelectric materials (polar but not switchable) from ferroelectric materials (polar and switchable). The feature to experimentally manipulate—to "write"—a material property that is connected to the order parameter extends the applicability of ferroic materials from merely passive devices like sensors to new fields such as actuators, memory cells or logic gates. While some of the established ferroic materials, in particular ferromagnets and ferroelectrics, are widely used in scientific and technological applications, ferrotoroidic materials are far from being technologically applicable. This is especially because the generation of an actual conjugate field seems elusive so far.
The following section starts with a discussion of the symmetry requirements to switch magneto-toroidal states together with experimental realisation possibilities. Then, I will show how to generate effective magnetic vortex fields in the macrospin-based nanomagnetic arrays on the unit-cell scale using the time-dependent magnetic field of a moving magnetic tip. We published these findings recently in NATURE NANOTECHNOLOGY [3]. Subsequently, I will shortly present unsuccessful poling attempts using external electric- and magnetic-field settings and I will discuss their main challenges. The chapter closes with the presentation of global and local toroidal-poling schemes in vortex-type nanomagnetic arrays.

6.1 Symmetry Analysis of Suitable Manipulation Schemes

As elaborated in Sect. 2.1.4, one essential requirement for poling a magneto-toroidal order is the application of a conjugate field $\vec{S}$ that simultaneously violates time-reversal ($\mathcal{T}$) and space-inversion ($\mathcal{I}$) symmetry. Despite providing a suitable symmetry, the field requires a microscopic coupling to the order. As already considered

J. Lehmann, *Toroidal Order in Magnetic Metamaterials*, Springer Theses,
https://doi.org/10.1007/978-3-030-85495-9_6

by several scientists, see Sect. 2.1.5, appropriately applied electromagnetic field configurations may couple to the toroidal order:

I: Since the magneto-toroidal state is based on a whirled magnetisation distribution, a magnetic vortex field ($\vec{S} \propto \vec{\nabla} \times \vec{H}$) at the level of a unit cell has been proposed to constitute the conjugate field [4, 5]. Macroscopically, time-reversal symmetry and space-inversion symmetry are broken due to its presence:

$$\begin{aligned} \mathcal{T}\vec{S} \propto \mathcal{T}(\vec{\nabla} \times \vec{H}) &= \vec{\nabla} \times (-\vec{H}) \propto -\vec{S}, \\ \mathcal{I}\vec{S} \propto \mathcal{I}(\vec{\nabla} \times \vec{H}) &= (-\vec{\nabla}) \times \vec{H} \propto -\vec{S}. \end{aligned} \tag{6.1}$$

Microscopically, such a field yields a Zeeman interaction of either sign due to the field coupling with the clockwise- or counter-clockwise-curled magnetisation state of a toroidal moment. Consequently the field reduces or increases the energy for one or the other toroidal domain state. The corresponding coercive field can be estimated by considering the local switching field of a single macrospin in the structure. In some cases, the coercive field may be reduced by a simultaneous application of a homogeneous magnetic bias field along the toroidal moment vector [6]. However, the experimental challenge lies in the generation of coherent circular fields of just the size of a unit cell [7, 8].

II: A momentum-like vector, as for instance an electric current parallel to the toroidal moment, has been proposed to be the conjugate field ($\vec{S} \propto \vec{j}_{\|}$) [4, 7, 9]. The electric current $\vec{j}$ is described as a charge q that moves with velocity $\vec{v} = \frac{\partial \vec{r}}{\partial t}$. Hence, time-reversal symmetry as well as space-inversion symmetry are broken macroscopically:

$$\begin{aligned} \mathcal{T}\vec{S} \propto \mathcal{T}(\vec{j}) &= q\,\frac{\partial \vec{r}}{-\partial t} \propto -\vec{S}, \\ \mathcal{I}\vec{S} \propto \mathcal{I}(\vec{j}) &= q\,\frac{-\partial \vec{r}}{\partial t} \propto -\vec{S}. \end{aligned} \tag{6.2}$$

This is expected since an electric current flowing through a system along the toroidal moment vector indeed generates a magnetic vortex field (see **I**) in the plane perpendicular to the current vector. Electric-current-based switching of a magnetically compensated order has been investigated successfully in some semimetallic materials [10, 11] in terms of spin-orbit-torque, staggered magnetic fields and the antiferromagnetic Edelstein effect. The relation of the latter with the switching of toroidal moments has been discussed very recently [12, 13]. With this approach, however, the manipulation of toroidal order in electrically insulating crystals can not be achieved [8].

Current-induced toroidal poling may be possible in my artificial nanomagnetic arrays by a minor reconsideration of the design. By guiding a current through the centre of each unit cell, for example by lithographically fabricated metallic connectors, a conjugate field may be generated for re-aligning the toroidal moment of individual unit cells on demand by changing the sign of the current flow. By lithographically contacting all of these 'current guides', electric field pulses may be used to achieve global toroidal poling.

III: Following Ampère's law in Maxwell's equations, an electric current can be replaced by a displacement current that equivalently generates a curled magnetic field ($\vec{S} \propto \frac{\partial \vec{E}_{\parallel}}{\partial t}$), see Ref. [4] p. 16f. Macroscopically, time-reversal symmetry and space-inversion symmetry are broken:

$$\begin{aligned} \mathcal{T}\vec{S} &\propto \mathcal{T}(\tfrac{\partial \vec{E}}{\partial t}) = \tfrac{\partial \vec{E}}{-\partial t} \propto -\vec{S}, \\ \mathcal{I}\vec{S} &\propto \mathcal{I}(\tfrac{\partial \vec{E}}{\partial t}) = \tfrac{-\partial \vec{E}}{\partial t} \propto -\vec{S}. \end{aligned} \tag{6.3}$$

Successfully exploiting this condition would constitute an ideal magnetoelectric device in which the toroidal order, as a robust magnetic order, can be re-aligned by ultrashort electric-field pulses, as for instance via a capacitor discharge.

IV: The linear magnetoelectric effect can be utilised as discussed in Sect. 2.1.6 on page 21. The simultaneous application of an electric field $\vec{E}_{\perp}$ to induce a magnetisation, and a magnetic field $\vec{H}_{\perp}$ to act on this induced magnetisation may result in a re-alignment of the toroidal state ($\vec{S} \propto \vec{E} \times \vec{H}$) [7, 14, 15]. This field-constellation macroscopically breaks time-reversal symmetry and space-inversion symmetry by one of the fields each:

$$\begin{aligned} \mathcal{T}\vec{S} &\propto \mathcal{T}(\vec{E} \times \vec{H}) = \vec{E} \times (-\vec{H}) \propto -\vec{S}, \\ \mathcal{I}\vec{S} &\propto \mathcal{I}(\vec{E} \times \vec{H}) = (-\vec{E}) \times \vec{H} \propto -\vec{S}. \end{aligned} \tag{6.4}$$

Since this conjugate-field configuration is actually composed of two separate fields, it is indistinguishable if the coupling to the system is of first order ($\vec{S} \sim \vec{T}$) or of second order ($(\vec{E}, \vec{H}) \sim \vec{T}$) and thus of ferromagnetoelectric nature, see Ref. [16] p. 4. Magnetoelectric switching has been achieved [15] in $LiCoPO_4$ with a coercive field of 250 TV/cm.

6.2 Poling of Macrospin-Based Toroidal Square Arrays

As mentioned in the previous section, a variety of poling schemes may be applicable for manipulating the toroidisation of a solid. Here, the use of artificial magneto-toroidal arrays opens new possibilities for the application of some of the elusive switching schemes. I will demonstrate in the next sections that an experimental realisation of toroidal switching at the mesoscale is possible by advanced techniques, yet remains an experimentally challenging task.

6.2.1 Magnetic-Tip-Induced Generation of Local Magnetic Vortex Fields

Motivated by in-situ imaging and manipulation during magnetic reversal processes, see Appendix C.1 on page 167, I developed a novel method for locally manipulating the toroidal state with an MFM tip. As already introduced in Sect. 4.2 on page 83, an area scan with a magnetic tip is not necessarily an interference-free observation, but the tip can influence the magnetic state of the sample. This is an undesired effect for imaging, but opens the door for local manipulation schemes. The effect of magnetostatic interactions between tip and sample has been analysed thoroughly [17–21]. The magnetic layer of a typical MFM tip generates a field of about 40 mT at a distance of 50 nm (10–20 mT at a distance of 100 nm) from the tip apex, see Ref. [18] p. 256.[1] A few very recent studies exist on the tip-assisted manipulation of magnetic states in artificial nanomagnetic arrays. Wang et al. applied a protocol for writing different (excited) states in arrays of artificial spin-ice structures. They demonstrated that an external field plus the MFM-tip field can be used to locally overcome the coercive field of a certain subset of nanomagnets in the structure [22]. Furthermore, Gartside et al. realigned the magnetisation in single isolated nanowires as well as nanowires within a kagome array (see Fig. 2.14b on page 36) using an MFM tip [41, 44, 45].

Here, I resorted to the magneto-toroidal square arrays, see Fig. 4.1a. By using an array with small separation distances in between building blocks, a magnetostatic coupling cannot be neglected. On the contrary, the short-range coupling between single macrospins is essential for the envisaged poling scheme. An array built from 20 nm thick aluminium-passivated permalloy nanomagnets with lateral dimensions of $450 \times 150\,\text{nm}^2$, a unit-cell size of 830 nm and macrospin-separation distances of 70 nm (edge to edge) has been used. The whole array has a size of $\sim 55 \times 55\,\mu\text{m}^2$. During magnetisation reversal, the preference for the ferrotoroidic state has been confirmed by locally resolving the magnetic configuration at the coercive field. I applied LMOKE to probe the in-plane net magnetisation M of the array as a function of applied magnetic field. As extracted from the measured hysteresis curve, see Fig. 6.1a, a magnetic field of ± 40 mT drives the array into a uniformly magnetised saturated state, spatially resolved in Fig. 6.1b. At the coercive field of about ± 27 mT, the net magnetisation vanishes and the sample enters a demagnetised low-energy state. This state turns out to exhibit a ferrotoroidic $\pm \vec{T}$ multi-domain pattern, see Fig. 6.1c, with no preference for one or the other direction of toroidisation. The toroidal order forms spontaneously during magnetic reversal as a consequence of the close coupling between macrospins.

The MFM-tip-assisted toroidal poling is based on the selective reorientation of individual nanomagnets. Again, a homogeneous magnetic field is applied to uniformly align all nanomagnets into a remanent state with the overall magnetisation pointing in one of the $\pm[110]$ directions. After this initiation of the state, no further application of external fields is needed, since the magnetic fields required for poling will be generated within the array, as explained later.

An MFM tip with high magnetic moment (PPP-MFMR, see Table 3.1) is used for the controlled manipulation. The tip is scanned with $10\,\mu\text{m/s}$ bidirectionally along the $\pm[110]$ 'fast' direction in semi-contact lift-height mode. The net motion of the tip is

[1] Further, the authors mention that "tip fields are sharply localised, and their effect cannot be guessed from that of an equivalent uniform field".

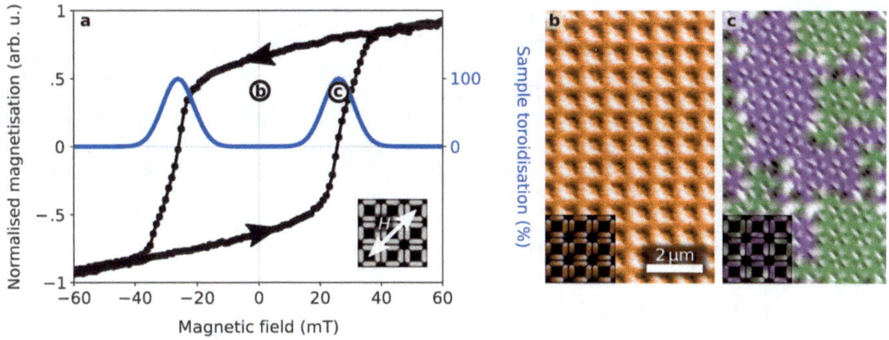

Fig. 6.1 Magnetisation curve and ferrotoroidic ground state. a In-plane magnetisation (M) curve of the toroidal square array measured with LMOKE as a function of magnetic field (H) along $\pm[110]$. Laser light of 405 nm wavelength, a power of $\sim$ 10 mW and a polarisation along $\pm[110]$ was used for the optical measurement, see Fig. 3.7. Blue curves represent Gaussian fits of $|\partial M/\partial H|$ that indicate the areal density of ferrotoroidic regions that peaks around the coercive magnetic field. **b** MFM image of the ferromagnetic single-domain state at remanence (orange: $\vec{M}_{[110]}$). The inset highlights the position of macrospins. **c** MFM image of the ferrotoroidic multi-domain state (violet: $-\vec{T}$, green: $+\vec{T}$) at the coercive magnetic field

in one of the $\pm[1\bar{1}0]$ 'slow' directions with 20 nm line-to-line distance. A small line-to-line spacing ensures to cross every nanomagnet of the array a few times. Whenever the tip passes across a nanomagnet, the nanomagnets internal energy landscape is perturbed by the field provided by the MFM-tip such that the magnetic state of the nanomagnet becomes destabilised. This process selectively lowers the local coercive field for the considered magnet. Concomitantly, the nanomagnet becomes susceptible to reorientation in the local magnetic-field distribution exerted by the neighbouring magnets.

In Fig. 6.2, four time steps τ_1 to τ_4 are depicted that are key for local toroidal poling of a representative unit cell defined by the four centred macrospins. The averaged magnetic field acting on a tip-destabilised nanomagnet, exerted by the surrounding magnets, is derived from a micromagnetic calculation, see Sect. 3.5, and is represented by a green block arrow and red contours. The calculation takes into account seven neighbouring macrospins that are positioned around the respective destabilised magnet, as shown in Fig. 6.2. The contribution of magnets which are even further away do not qualitatively change the result and are therefore neglected. On the contrary, even the consideration of just three (one parallel and two perpendicular) neighbours does not lead to qualitative changes of the resulting net field, see Appendix C.2. The calculations do not include the magnetic field of the MFM tip, since it enables the reorientation process but does not determine the newly set direction of magnetisation.[2] Following the trajectory of the tip, all four nanomagnets of a unit cell are

[2] Using a model system of anisotropic (Ising-like) point-dipoles arranged on a square lattice, Dr. Peter Derlet from the Condensed Matter Theory Group at Paul Scherrer Institute examined the existence of a perturbation-regime where the strength of the tip field only destabilises the

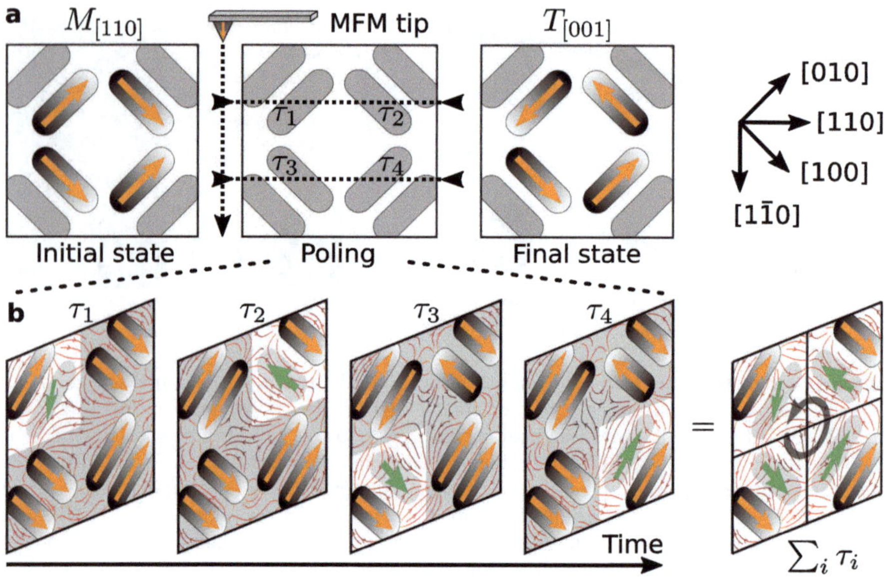

Fig. 6.2 **Local toroidal poling via crystal magnetic fields.** **a** Poling of a toroidal square array. The schematic MFM scan over the unit cell in its initial state with a net magnetisation $\vec{M}$ along [110] yields a toroidal state with a toroidisation $\vec{T}$ along [001]. The sketch of the MFM scan in the centred panel with the tip trajectory indicated by the dashed line is for representative purposes only, since the actual bidirectional raster scan has a much smaller line-to-line spacing. **b** Sequential tip-assisted 'activation' and reorientation of the four nanomagnets of a unit cell. The magnetic field exerted by the magnetic moments (orange block arrows) is indicated by red contours (see Fig. 3.12). The green block arrows represent the calculated direction of the net magnetic field averaged over the area of the respective nanomagnet above which the tip hovers at time τ_i. The direction of this net magnetic field determines whether the nanomagnet will reverse its magnetisation (τ_1, τ_2) or remain in its initial magnetic state (τ_3, τ_4). The combination of the four time steps (bottom-right) yields an effective magnetic vortex field (black circular arrow) at the level of the unit cell

destabilised in a sequence at the four time steps ($\tau_1 - \tau_4$), see Fig. 6.2b. The field acting on the respective destabilised macrospin is determined as an average (green block arrows in Fig. 6.2b) of the magnetic field distribution calculated from equation 3.11 (small red contour arrows in Fig. 6.2b) across the area of this particular nanomagnet. This net field determines the new direction of magnetisation for the considered nanomagnet. Sequentially combining time steps τ_1 to τ_4, as shown at the bottom-right of Fig. 6.2b, reveals the generated effective magnetic vortex field. The curl of this effective magnetic-field distribution that is generated at the level of a single unit cell, $\vec{S}_{\text{eff}} \equiv \vec{\nabla} \times \vec{H}$, can be calculated in a discrete form. By considering a planar field distribution $\vec{H} = (H_x, H_y, 0)$ and zero derivation of the field along z ($\partial/\partial z = 0$), the two-dimensional expression can be written by using *central differences* [24] as

---

orientation of individual nanomagnets. Therefore it only catalyses reorientation processes and does not contribute to the switching of single nanomagnets as such [23].

$$\vec{\nabla} \times \vec{H} = \left[\left(\frac{H_{4,y}(x + |r|) - H_{1,y}(x - |r|)}{2\,|r|} \right) - \left(\frac{H_{2,x}(y + |r|) - H_{3,x}(y - |r|)}{2\,|r|} \right) \right] \vec{e}_z \,, \tag{6.5}$$

with $H_{i,j}$ being the magnetic field at the position of the i-th magnet along the $j = \{x, y\}$ direction and r the magnet's distance from the centre of the unit cell. For the generation of such a local magnetic vortex field, the interaction strength that couples neighbouring magnets is of significant importance. In particular the field that is acting on the two nanomagnets at time steps τ_3 and τ_4 depends drastically on the precise coupling energies between perpendicular and parallel magnets ($E_\perp$ and $E_\parallel$, see Fig. 3.11). At these time steps, the interaction with two perpendicular neighbours ($E_\perp$) prohibits a re-orientation of the considered magnet, while the interaction with the one parallel neighbour ($E_\parallel$) favours a re-orientation. Since no re-orientation is desired at time steps τ_3 and τ_4 for the generation of a magnetic vortex field, the relation $E_\perp > E_\parallel/2$ has to be satisfied (Results from this calculation are shown in Appendix C.2).

Note that the effect on the unit-cell level can be transferred to a scan of the entire array and therefore yields a time-integrated global toroidal field that is acting on the area scanned by the magnetic tip. The micromagnetic calculations and, subsequently, Eq. 6.5, can be calculated for both slow scan directions that lay perpendicular to the prepared direction of net magnetisation. The result is a scan-direction-dependent sign of the effective magnetic vortex field acting on the array (as experimentally verified, see Fig. 6.3). As a consequence, the reversal of the scan direction allows for a poling into the oppositely oriented toroidal state.

For scrutinising this intriguing toroidal-poling scheme experimentally, different scan parameters have been applied across four $12 \times 12\,\mu\text{m}^2$-sized areas of the array, as shown in Fig. 6.3. Apart from a few local misaligned macrospins, the four areas are homogeneously poled in either the $-\vec{T}$ or $+\vec{T}$ orientations. As demanded by symmetry, the reversal of the slow scan direction (Fig. 6.3a,b $\rightarrow$ c,d) introduces a time-reversal operation into the poling process. Consequently, the reversal leads to a reversal of the sign of the effective toroidal field and thus of the resulting toroidisation. Further, a reversal of the initial magnetisation direction of the sample is equivalent to a sample rotation by 180° around its surface normal, and therefore equals the aforementioned reversal of the scanning direction.

The experiment confirms a negligible contribution of the tip-magnetisation direction upon the generation of a toroidal poling field. A reversal of the tip magnetisation does not affect the resulting toroidisation and corroborates the suggestion that the MFM-tip field merely 'catalyses' the process. To reveal the applied poling schemes without further altering the magnetic configuration, the array has been scanned with a tip of low magnetic moment (PPP-LM-MFMR, see Table 3.1) using a lift-height of 45 nm.

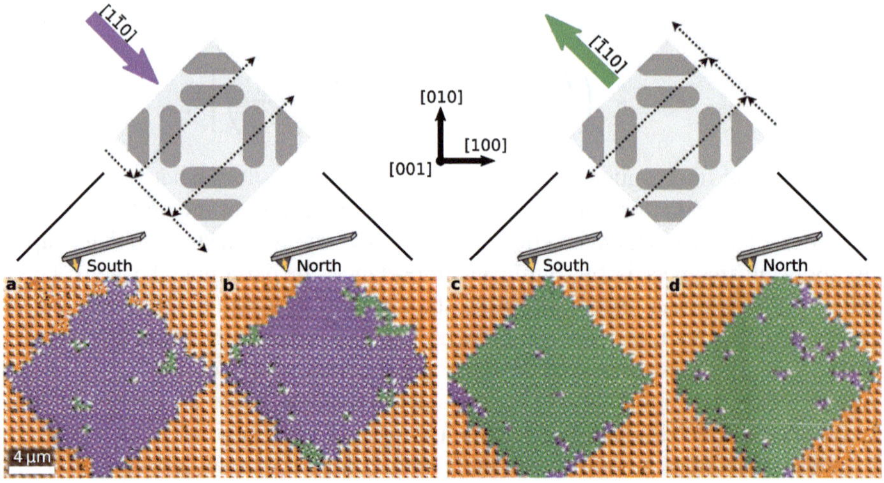

Fig. 6.3 Experimental demonstration of local magneto-toroidal poling. a–d MFM images of four regions of the toroidal square array to which toroidal poling fields $+\vec{S}$ or $-\vec{S}$ were applied in $12 \times 12\,\mu\text{m}^2$-sized areas. The array is built from nanomagnets of size $450 \times 150 \times 20\,\text{nm}^3$ with a unit-cell size of 830 nm and macrospin-separation distances of 70 nm (edge to edge). The initial state of the array exhibits a net magnetisation along $-[110]$ (orange-coloured area). Different combinations of slow scan directions (violet and green block arrows) and out-of-plane tip magnetisations (North and South) are schematically shown above the scans. The sign of the toroidal poling field ($-\vec{S}$ in **a**, **b**; $+\vec{S}$ in **c**, **d**, see Fig. 6.2) determines the resulting toroidisation (violet: $-\vec{T}$, green: $+\vec{T}$) and depends on the slow-scan direction but not on the magnetisation direction of the tip. For more insight, the Appendix Fig. C.2 displays the MFM scan of the entire array together with an 'in-situ' imaging of the switching events during the four scans

6.2.2 *Application of Displacement Currents by Electric Field Gradients*

As reviewed above, a polar c-type vector such as an electric current may act as a conjugate field and therefore suit the purpose of toroidal switching. With Maxwell's arrangement of the fundamental theory of electrodynamics it is known that a current does not necessarily have to consist of a flow of delocalised charge carriers, but that a displacement current ($\vec{j}_\text{d} = \epsilon_0 \epsilon_\text{r} \frac{\partial \vec{E}}{\partial t}$) serves as an analogous representation with the same symmetry. As a textbook example, a displacement current is present when capacitors are being charged or discharged or—equivalently—whenever electric fields build up or decay. Since the alternating build-up and decay of electric (and magnetic) fields is inherent to any electromagnetic wave, light may be an ideal tool for poling toroidal order. Yet, it is crucial to consider that the field and its derivative change sign every half-cycle of the oscillation. Thus, the net effect of such an alternating displacement current is zero ($\int_{-\infty}^{\infty} E(t)dt = 0$, with electric field $E(t)$ and time t).

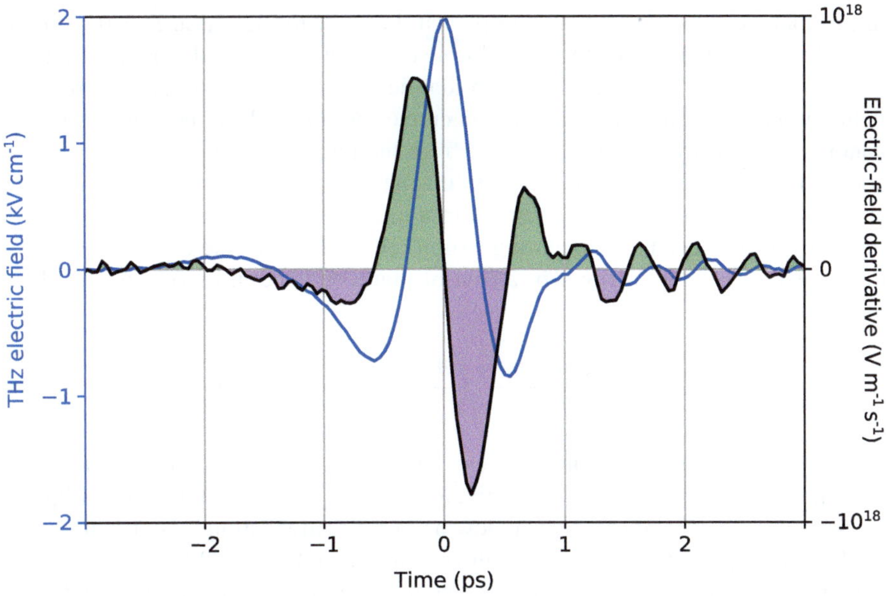

Fig. 6.4 Displacement current as a conjugate toroidal field. Electric-field profile of a single-cycle terahertz pulse (blue). Its time derivative is proportional to the effective displacement current in vacuum (black) and shows a distinct maximum followed by a distinct minimum. The respective displacement current constitutes an alternating sequence of conjugate-field pulses of both negative ($-\vec{S}$, violet) and positive ($+\vec{S}$, green) sign

The generation of coherent and phase-stabilised ultrashort pulses of electromagnetic radiation, however, allows to circumvent this issue. A few-cycle electromagnetic pulse provides a time-wise highly confined electric field that yields a displacement-current function with only a few minima and maxima. Each of these peaks lying above the coercive field may induce a switch. The final state will therefore be defined by the last extrema of the displacement-current function. As a result, a few-cycle field pulse with controllable carrier-envelope phase possibly allows for a toroidal poling with either sign.

The TERADYNAMICS sub-group in our research group supervised by Dr. Shovon Pal uses ultra-short few-cycle terahertz pulses with electric-field amplitudes of up to 2 kV/cm for probing excitations in strongly correlated materials. Such a field pulse together with its time derivative is shown in Fig. 6.4.

Terahertz field pulses have been applied to the sample in grazing incidence with an angle of approximately 85° to the surface normal. The linear polarisation of the field was parallel to the plane of incidence to ensure an electric-field derivative almost parallel to the out-of-plane laying toroidisation vector. A toroidal square array, see Fig. 4.1a, composed of 450 × 150 × 20 nm^3-sized nanomagnets with a lattice constant of 830 and 70 nm edge-to-edge distances has been used. A homogeneous magnetic field above the coercive field of the array along [110] defined the

initial homogeneously magnetised remanent state with toroidisation $\vec{T} = 0$. Thus, all deviations from the initial magnetised state can be attributed to the poling attempt.
An approximation for the coercive toroidal field is needed to estimate the required field-derivative amplitude. Since the coercive toroidal field is not known for the nanomagnetic array, it may be derived from the array's coercive magnetic field of about 30 mT, see Fig. 6.1. To generate magnetic fields in the order of the 30 mT at the position of the four macrospins in a unit cell, a calculation of the required (displacement) current density through a unit cell can be performed. The local tangent magnetic flux density generated by an out-of-plane-oriented current can be calculated as

$$\vec{B} = \frac{\mu_0\mu_r I_z}{2\pi r}\vec{e}_\theta = \frac{\mu_0\mu_r\epsilon_0\epsilon_r l_c^2}{2\pi l_c/2}\left(\frac{\partial E_z}{\partial t}\right)\vec{e}_\theta = \frac{\mu_r\epsilon_r l_c}{\pi c^2}\left(\frac{\partial E_z}{\partial t}\right)\vec{e}_\theta, \quad (6.6)$$

with ϵ_0 and μ_0 (ϵ_r and μ_r) being the permittivity and permeability of vacuum (in the material under consideration), respectively and $c = \sqrt{\epsilon_0\mu_0}^{-1}$ is the speed of light in vacuum. The (displacement) current flowing in z-direction through a unit cell is given by $I_z = \epsilon_0\epsilon_r\frac{\partial E_z}{\partial t}l_c^2$ with l_c being the lattice constant and $r \approx l_c/2$ being the distance from the centre of a unit cell to its four comprising macrospins. Consequently, the generation of a 30-mT-large magnetic field at the position of the macrospins requires a electric-field derivative of the pulse of

$$\left(\frac{\partial E_z}{\partial t}\right) \geq \frac{\pi B c^2}{\mu_r\epsilon_r l_c} \approx \frac{1}{\mu_r\epsilon_r}10^{24}\,\text{V/m s} \quad . \quad (6.7)$$

Considering the literature values for the relative dielectric permittivity of silicon [25] $\epsilon_r^{\text{Si}} = 11.7$ (the substrate material) and for the relative magnetic permeability of permalloy [26], $\mu_r^{\text{Py}} = 10^4$ (the alloy the macrospins are made of), an electric-field derivative of $\partial E/\partial t \gtrsim 10^{19}$ V/m s is required for toroidal poling. While the required electric field is about an order of magnitude larger than the actual electric field of the THz pulse, see Fig. 6.4, it might be still sufficiently large to overcome the switching barriers of the sample. This is because the relative permittivity and permeability of silicon and permalloy, respectively, may be larger than the assumed DC values close to resonances located in the terahertz-frequency range.
I performed two poling experiments using terahertz-pulse trains incident from either of the two sides of the sample. For each setting, an illumination time of 1 s has been chosen, such that, with a pulse repetition rate of 1 kHz, about a thousand pulses reached the sample. To study the affect of the exposure, the magnetic state of the sample has been measured subsequently using MFM. Unfortunately, after both experiments, no deviation from the initially set remanent magnetised state of the array has been detected.
For consecutive experiments, the use of very thin nanomagnets would be beneficial, since the switching barrier and with it the coercive magnetic field scales linear with the nanomagnet's thickness. In addition, a substrate with a large relative dielectric permittivity ϵ_r, for instance a ferroelectric material such as $BaTiO_3$, would lead

to an increased displacement current. Most important, the time scales of the field application have to be reconsidered. Since my nanomagnetic arrays comprise of nanometre-sized ferromagnetic building blocks, each building block may not be able to respond on picosecond time scales (which is the typical time scale for the description of the dynamics in the spin system of magnetically-compensated ordered materials [27, 28]) but rather on nanosecond time scales, governed by corresponding ferromagnetic-resonance frequencies [29, 30]. Accordingly, the pulse-duration time of a few picoseconds is presumably below the precession and dissipation times for the spins in permalloy nanostructures, which prevents a coherent reorientation upon terahertz excitation. In order to corroborate suitable time scales for interacting with the nanomagnetic toroidal arrays, simulations based on the Landau-Lifshitz-Gilbert equation [31, 32] may be required.

6.2.3 Application of Crossed Electric and Magnetic Fields

Poling of the toroidal order in crossed static electric and magnetic fields has been demonstrated on electrically insulating ferrotoroidic crystals [15, 33]. An adaption of this DC poling scheme for nanomagnetic arrays may, however, be impeded by the conducting and ferromagnetic character of permalloy as the system's building-block material. Static electric fields would get screened and any effect possibly induced by the field would be nullified. Further, static magnetic fields with an amplitude above the coercive magnetic field of a nanomagnet would induce a remanent magnetised state with zero toroidisation. Both obstacles may be addressed by resorting to AC fields such that the electric-field-screening efficiency inside the material can be attenuated. Hence, magnetic-field amplitudes that just not perturb of the magnetic order and electric fields sufficiently low to avoid an electrical discharge appear reasonable. While both fields average out over time, the simultaneous application of AC electric and magnetic fields that oscillate with a 0° or 180° phase against each other yield a non-zero conjugate field, see Fig. 6.5.

Electric and magnetic fields have been applied in the plane ($\vec{E} \parallel [1\bar{1}0]$ and $\vec{H} \parallel [110]$) of a macrospin-based toroidal square array, see Fig. 4.1a, composed of $450 \times 150 \times 20\,\text{nm}^3$-sized nanomagnets, arranged with a lattice constant of 830 nm and 70 nm edge-to-edge distances between the magnets. A remanent magnetised state has been initiated using a homogeneous magnetic field above the coercive field of the nanomagnetic array along the [110]-direction. Hence, all deviations from this pre-set single-domain state can be attributed to changes induced by the applied field. In-plane in- or anti-phased electric and magnetic AC fields have been generated by two approaches. A low-frequency setup involves a modified high-voltage stray-field transformer operating at a frequency of 50 Hz (see Appendix B.4) that generates a magnetic field up to 120 mT and an electric field up to 6 kV (with electrode distances of 15 mm). A high-frequency setup involves a zero-voltage-switching-based resonance-transformer circuit operating at 300 kHz (see Appendix B.4) that gener-

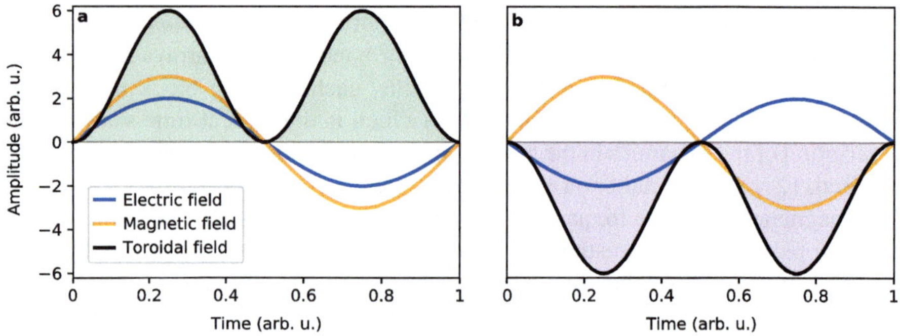

Fig. 6.5 Magnetoelectric generation of a conjugate field. The concept to obtain an effective conjugate field ($\vec{S} = \vec{E} \times \vec{H}$) utilises 0° or 180° phase-locked oscillating electric ($\vec{E}$) and magnetic ($\vec{H}$) fields. Both fields average out over time, yet their cross product yields a finite averaged amplitude with well-defined sign. **a** In-phase AC fields generate a positive conjugate field $+\vec{S}$ (green shading) perpendicular to $\vec{E}$ and $\vec{H}$. **b** Anti-phase AC fields generate a negative conjugate field $-\vec{S}$ (violet shading) perpendicular to $\vec{E}$ and $\vec{H}$

ates a magnetic field of about 10 mT and an electric field of about 1 kV. The magnetic field is generated with an induction coil that is placed on a soft-magnetic ring-like bore with a gap at the sample position. The electric field is generated via induction using a secondary coil placed on the same bore. This generation scheme ensures phase-synchronised operation by default.

By applying oscillating magnetic fields but no electric fields, it was verified that the magnetic field alone is not able to influence the pre-set remanent magnetised state of the array, which has been confirmed via MFM. Subsequent poling tests have been performed with both fields applied for a few seconds. The array has been probed via MFM to reveal possible changes in the magnetic configuration after each test. The process cycle of manipulating the sample in the poling device and consecutively measuring the magnetic configuration in the MFM device is time ineffective and laborious. Therefore, and due to the unknown coupling constant of the fields to the toroidal state, only a few poling tests have been performed using the maximum reachable electric-field strengths generated by the devices. However, no change of the magnetic configuration has been uncovered after the application of crossed AC electric and magnetic fields. I could not determine whether this is because the coercive field has not been reached, the oscillation frequencies were too low, or the coupling of the fields to the structure is too small. In order to address this uncertainty, a method for in-situ monitoring of switching events is required.

6.3 Poling of Vortex-Type Magneto-Toroidal Arrays

As compared with the demonstrated tip-assisted toroidal poling of macrospin-based nanomagnetic arrays, the toroidal poling of arrays based on vortex elements functions differently on the microscopic scale. It has been shown that the vortex state in ferromagnetic elements that structurally violate space-inversion symmetry can be reversed by the application of a homogeneous magnetic field [34–38]. In contrast to the structurally centrosymmetric macrospin-based arrays (see Fig. 6.1), either of the two clockwise- and counter-clockwise-whirled vortex states (corresponding to a $-\vec{t}$ and $+\vec{t}$ toroidal moment, respectively) can be accessed directly during the magnetisation reversal, see Fig. 5.13 on page 107. This is possible due to a shape-dependent asymmetric nucleation energy of the vortex core: For the case of a triangle-shaped nanomagnetic element, the energy barrier for the formation of a vortex core at a corner is higher than its formation at an edge. Since the location of nucleation with respect to the magnetisation direction determines the resulting sense of vorticity, the corresponding toroidal moment is linked one-to-one to either the positive or the negative coercive magnetic field.

For controlling the net toroidisation of a certain crystal just with a homogeneous magnetic field (breaking the time-reversal symmetry), its structure has to exhibit a polar axis, which breaks space-inversion symmetry. The concept of a polar axis in nanostructures and a resulting 'virtual polarisation' has been introduced and used in Ref. [39] to express the particular ferroic symmetry of 'v'-shaped nanomagnetic elements. As discussed in the beginning of this chapter, this time- and space-odd symmetry is required to manipulate the ferrotoroidic state. For this particular case, a linear magnetoelectric interaction [40] can be assumed with a conjugate field that is composed of a magnetic and—perpendicular to it—a 'virtual electric field'. Indeed, the sign of this toroidal field (and the resulting toroidisation of the sample) depends on two variables that parametrise the broken space-inversion and time-reversal symmetry: The direction of 'virtual polarisation' and the direction of the magnetisation in the pre-set magnetised state.

With respect to the potentially ferrotoroidic arrays, highlighted in Fig. 4.1d–f, the introduced symmetry requirement is fulfilled only for array (d). Figure 6.6 reveals a uniform toroidal state in this type of array, measured after approaching its positive coercive field (plateau region in Fig. 5.13) from a state of magnetic saturation along $[\bar{1}00]$.

Besides this global poling scheme in a homogeneous external magnetic field, a tip-assisted toroidal poling for achieving local control over the order parameter, see Sect. 6.2.1, is applicable as well. The same array as used for global toroidal poling has been used (see Fig. 4.1d, 400 nm edge length, 20 nm thickness, 50 nm corner radii and 100 nm triangle separations). An initial state with uniform magnetisation along the triangles edges is defined with an external magnetic field above the coercive field of the array along [100]. By subsequently scanning the array with a tip of high magnetic moment (PPP-MFMR, see Table 3.1), as discussed in Sect. 6.2.1, the magnetic state of each triangle within the scanned area can be re-configured. The out-of-plane-magnetised MFM tip perturbs the magnetic state via its magnetic stray fields. Hence, at the first contact with a triangle, the tip field pulls a magnetic vortex [41] into the triangle and herewith defines both, the magnetisation direction of the vortex core as well as the vorticity and accordingly the toroidal moment. As a result, the scanned triangle undergoes a field-induced phase transition from an excited (magnetised) state into the toroidal ground state. The pathway for this transition and, associated with

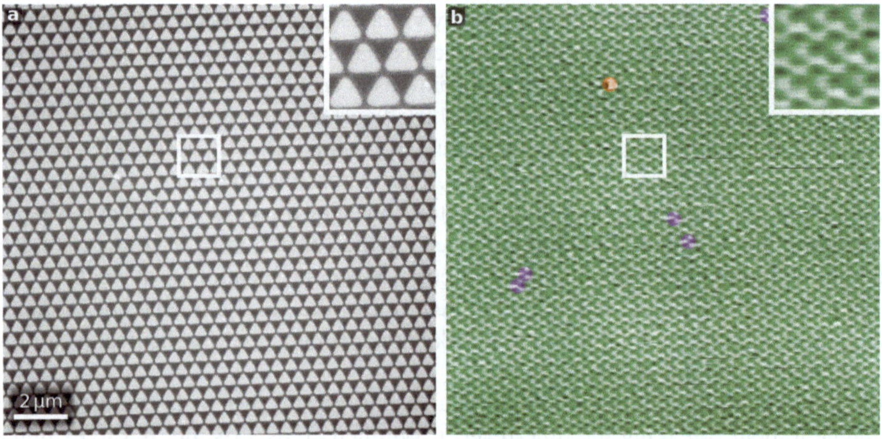

Fig. 6.6 Toroidal poling of a non-centrosymmetric array of ferromagnetic triangles. With the reversal of a uniform magnetic field that is aligned horizontally, and thus, parallel to the triangle's edge, the system can be poled into either of the two toroidisation states $\pm\vec{T}$, see Fig. 5.13a and c, respectively. The array comprises permalloy triangles with 400 nm edge length, 20 nm thickness, 100 nm element-to-element separation and 50 nm corner radii. **a** The topography revealed with AFM. **b** The magnetic configuration revealed with MFM. Here, from saturation along [$\bar{1}$00] the positive coercive field has been approached (field along the right, [100] direction) in order to induce a uniform toroidal state of positive sign (violet: $-\vec{T}$, green: $+\vec{T}$)

it, the particular realisation of the final state is, however, not random, but controlled by the scan direction of the magnetic tip, see Fig. 6.7. The measurement confirms that the shape-determined bias for the location of the vortex-core nucleation can be overcome by the tip trajectory that finally controls the position of nucleation.

The comparison of local toroidal poling in macrospin-based arrays (Sect. 6.2.1) and vortex arrays, as discussed here, leads to several striking conclusions. Remember that, the poling mechanism in macrospin-based arrays requires a strong coupling between the nanomagnets, which acts as the source for the effective poling field. In contrast, the poling in vortex-based arrays is applicable for isolated toroidal moments as well and does not necessitates a coupling. Hence, the poling mechanism in vortex-based structures appears to be understood as a controlled relaxation scheme in which the travelling direction of the magnetic-tip field determines the resulting ground state. The equivalence of both systems with a final state of well-controlled net toroidisation after the application of the poling protocol is remarkable and provides motivation for a generalised theoretical description of the poling mechanisms. Such a description is further motivated with the experimental conditions that comply with the violation of time-reversal and space-inversion symmetry required for the generation of a conjugate field to the toroidal state.

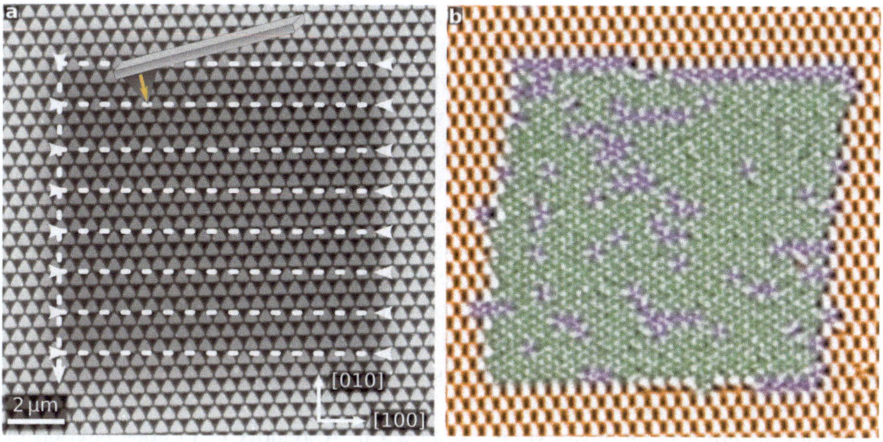

Fig. 6.7 Magnetic-tip-assisted local toroidal poling of an array of ferromagnetic triangles. A raster scan across the nanomagnetic array with an MFM tip of high magnetic moment induces local switching processes. Therefore, the array has been initiated into a remanent magnetised state along [100]. **a** Topography of the array, measured via AFM. The scan over an area of $12 \times 12\,\mu m^2$ (slow scan direction along [0$\bar{1}$0]) is indicated with dotted lines. **b** A scan with a tip of low magnetic moment revealed the effect of the poling procedure. The scanned area has been converted from a net-magnetised state (orange: $\vec{M}_{[100]}$) into a state with a positive net toroidisation (violet: $-\vec{T}$, green: $+\vec{T}$)

6.4 Summary and Perspective

Conjugate-field poling constitutes one of the hallmarks of a ferroic state. In zero external field, the possible domain states in a ferroic material, see Chap. 5, are energetically degenerate, i.e., they collectively align yet without a preferential direction. Conjugate-field poling describes the reversal of one or more domain-orientations in favour of another orientation due to a field-induced imbalance of the internal energy. Since the controlled manipulation of ferroic domain states is essential for most of the applications, it is of key importance to understand poling mechanisms and to elaborate experimentally feasible poling schemes. For the widely accepted ferroic states, ferromagnetism, ferroelectricity and ferroelasticity, the conjugate fields are the magnetic field, the electric field, and mechanical stress, respectively. The application of any of these fields is experimentally well established. In contrast, the poling of ferrotoroidic order appears fundamentally challenging. Despite the theoretically considered options for a conjugate toroidal field, experimental realisations appear elusive.

I introduced different approaches for generating a space-inversion- and time-reversal-symmetry-breaking field, supported by an evaluation of their microscopic-coupling mechanism with the toroidal state. Further, I experimentally demonstrated a local toroidal-poling scheme that is based on the creation of effective magnetic vortex fields of either sign interacting with the structure at the unit-cell level. Here, the transfer to mesoscopic length scales and the use of coupled artificial nanomagnetic arrays has proved particularly beneficial due to the direct spatial access to the magnetic configuration of the material. The applied poling procedure sequentially generates the required fields by scanning the array with a magnetic tip. The stray field of the magnetic tip destabilises the local magnetisation distribution in single nanomagnets and 'catalyses' unit-cell-scale magnetic vortex fields within the scanned area of the array. As a result, the nanomagnets sequentially sense the superimposed magnetic field exerted by the surrounding nanomagnets and can switch. Thus, local control over the orientation of the toroidal moment has been achieved. Note however, that a transition from one domain state to the other requires the field-induced generation of a remanent state with net magnetisation that sets the initial conditions for the poling protocol. Consequently, the initial state with net magnetisation is a necessary ingredient of the composed poling field. The polability of toroidal states of either sign (and different magnetised state, as shown in Appendix C.1) goes beyond the work presented by Wang et al. [22] where the combination of the external field and the tip field has been used for influencing the configuration in nanomagnetic spin-ice arrays.

My result constitutes a realisation of the conceptual staggered-field-switching scheme conceived by Jungwirth et al. for poling of a magnetically compensated order [42]. The ability to access such exotic field configurations offers new pathways for the manipulation of and control over spin states in other types of magnetically ordered states. Further work might be to investigate whether it is possible to apply the tip-assisted toroidal poling procedure to a toroidally ordered multi-domain structure. Here, the question would be if the protocol is applicable in general to excited states of the system, particularly to domain-wall states, in order to move a domain wall across the array.

Two additional poling protocols have been conceived and tested that involve the application of displacement currents, on the one hand, and of crossed electric and magnetic fields on the other hand. Although both methods potentially provide a conjugate field, either turned out to be experimentally infeasible. The main reason for this is possibly the application of displacement currents at timescales below intrinsic spin-relaxation times as well as field strengths in either of the two methods below the coercive field. The laborious two-step process of applying the respective protocol and subsequently measuring possible changes by MFM hindered a comprehensive examination of a more complete range of times scales and field strengths. A promising approach for future poling experiments could be the application of these conjugate fields during the system's transition into the ferroic phase. Such a field-cooling procedure enhances the susceptibility of the system upon field application, yet it would require very thin nanomagnetic elements or a substitution of the ferromagnetic material (for example with FeRh [43]) for lowering the transition temperature towards more accessible temperatures.

For arrays built from triangle-shaped nanomagnets, ensuing theoretical work may answer the question if the poling in either a uniform magnetic field or via a tip-assisted scanning process can be described in terms of the application of a conjugate field. In both cases, my experimental results confirmed the manipulation of the array into a single-domain toroidal state and therefore indicate that a conjugate field has probably been applied. However, the nature and microscopic key components of that field are not entirely understood yet.

As a side result from the tip-assisted toroidal poling of magnetic-vortex arrays, the applied protocol might act on not only the vorticity of the induced magnetic-vortex states but also onto the out-of-plane magnetisation direction of the vortex-core as a second degree of freedom. Since the magnetic tip provides a magnetic field that are aligned either parallel or antiparallel to the sample's surface normal, the magnetic vortex state of each triangle will be accompanied by a vortex-core-magnetisation direction determined by the tip-magnetisation direction.

References

1. Wadhawan VK (200) Introduction to ferroic materials, 1st edn. CRC Press. http://dx.doi.org/10.1201/9781482283051
2. Tagantsev AK, Cross LE, Fousek J (2010) Domains in ferroic crystals and thin films. Springer, New York. http://dx.doi.org/10.1007/978-1-4419-1417-0
3. Lehmann J et al (2019) Poling of an artificial magneto-toroidal crystal. Nature Nanotechnol 14(2):141–144. https://doi.org/10.1107/S1600576719001948
4. Dubovik V, Tugushev V (1990) Toroid moments in electrodynamics and solid-state physics. Phys Rep 187(4):145–202. https://doi.org/10.1016/0370-1573(90)90042-Z
5. Gorbatsevich AA, Kopaev YV (1994) Toroidal order in crystals. Ferroelectrics 161(1):321–334. https://doi.org/10.1080/00150199408213381
6. Dubovik V, Martsenuyk M, Martsenuyk N (1995) Reversal of magnetization of aggregates of magnetic particles by a vorticity field and use of toroidness for recording information. J Magn Mag Mater 145(1–2):211–230. https://doi.org/10.1016/0304-8853(94)01594-5
7. Schmid H (2001) On ferrotoroidics and electrotoroidic, magnetotoroidic and piezotoroidic effects. Ferroelectrics 252(1):41–50. https://doi.org/10.1088/0953-8984/20/43/434201
8. Spaldin NA, Fiebig M, Mostovoy M (2008) The toroidal moment in condensed-matter physics and its relation to the magnetoelectric effect. J Phys Condens Matter 20(43). https://doi.org/10.1103/PhysRevB.88.094429
9. Artamonov YA, Gorbatsevich AA, Kopaev YV (1984) Magnetoferroelectric and toroidal ordering. J Exp Theor Phys 40(7):290–293
10. Wadley P et al (2016) Electrical switching of an antiferromagnet. Science 351(6273):587–590. https://doi.org/10.1126/science.aab1031
11. Baltz V et al (2018) Antiferromagnetic spintronics. Rev Mod Phys 90(1). https://doi.org/10.1103/RevModPhys.90.015005
12. Watanabe H, Yanase Y (2018) Symmetry analysis of current-induced switching of antiferromagnets. Phys Rev B 98(22). https://doi.org/10.1103/PhysRevB.98.220412
13. Thöle F, Keliri A, Spaldin NA (2020) Concepts from the linear magnetoelectric effect that might be useful for antiferromagnetic spintronics. J Appl Phys 127(21). https://doi.org/10.1063/5.0006071
14. Schmid H (2008) Some symmetry aspects of ferroics and single phase multiferroics. J Phys Condens Matter 20(43). https://doi.org/10.1088/0953-8984/20/43/434201
15. Zimmermann AS, Meier D, Fiebig M (2014) Ferroic nature of magnetic toroidal order. Nat Commun 5(1):4796. https://doi.org/10.1038/ncomms5796
16. Litvin DB (2008) Ferroic classifications extended to ferrotoroidic crystals. Acta Crystallogr Sect A Found Crystallogr 64(2):316–320. https://doi.org/10.1107/S0108767307068262

17. Kleiber M et al (1998) Magnetization switching of submicrometer Co dots induced by a magnetic force microscope tip. Phys Rev B 58(9):5563–5567. https://doi.org/10.1103/PhysRevB.58.5563
18. Hopster H et al (eds) (2005) Magnetic microscopy of nanostructures. Springer, Berlin
19. Mironov V et al (2007) MFM probe control of magnetic vortex chirality in elliptical Co nanoparticles. J Magn Magn Mater 312(1):153–157. https://doi.org/10.1016/j.jmmm.2006.09.032
20. Garcia R, Knoll AW, Riedo E (2014) Advanced scanning probe lithography. Nat Nanotechnol 9(8):577–587. https://doi.org/10.1038/nnano.2014.157
21. Kazakova O et al (2019) Frontiers of magnetic force microscopy. J Appl Phys 125(6). https://doi.org/10.1063/1.5050712
22. Wang Y-L et al (2016) Rewritable artificial magnetic charge ice. Science 352(6288):962–966. https://doi.org/10.1126/science.aad8037
23. Derlet PM. Private communication. (unpublished data available)
24. https://scicomp.stackexchange.com/questions/21915/discrete-definitionsof-curl-nabla-times-f. (Online content, Accessed on 15 Jan 2020)
25. Afsar MN, Chi H (1994) Millimeter wave complex refractive index, complex dielectric permittivity and loss tangent of extra high purity and compensated silicon. Int J Infrared Millimeter Waves 15(7):1181–1188. https://doi.org/10.1007/BF02096073
26. Jiles D (2015) Introduction to magnetism and magnetic materials, 3rd edn. CRC Press. https://doi.org/10.1201/b18948
27. Gomonay EV, Loktev VM (2014) Spintronics of antiferromagnetic systems (Review Article). Low Temp Phys 40(1):17–35. http://aip.scitation.org/doi/10.1063/1.4862467
28. Sander D et al (2017) The 2017 magnetism roadmap. J Phys D Appl Phys 50(36). https://doi.org/10.1088/1361-6463/aa81a1
29. Bang W et al (2019) Ferromagnetic resonance spectra of permalloy nano-ellipses as building blocks for complex magnonic lattices. J Appl Phys 126(20):203902. http://aip.scitation.org/doi/10.1063/1.5126679
30. Lendinez S, Jungfleisch MB (2020) Magnetization dynamics in artificial spin ice. J Phys Condens Matter 32(1). https://doi.org/10.1088/1361-648X/ab3e78
31. Landau L, Lifshitz E (1935) On the theory of the dispersion of magnetic permeability in ferromagnetic bodies. Physikalische Zeitschrift der Sowjetunion, pp. 153–169. http://dx.doi.org/10.1016/B978-0-08-036364-6.50008-9
32. Brown WF (1959) Micromagnetics, domains, and resonance. J Appl Phys 30(4):S62–S69. https://doi.org/10.1063/1.2185970
33. Baum M et al (2013) Controlling toroidal moments by crossed electric and magnetic fields. Phys Rev B 88(2). https://doi.org/10.1103/PhysRevB.88.024414
34. Jaafar M et al (2008) Field induced vortex dynamics in magnetic Ni nanotriangles. Nanotechnology 19(28). https://doi.org/10.1088/0957-4484/19/28/285717
35. Jaafar M et al (2010) Control of the chirality and polarity of magnetic vortices in triangular nanodots. Phys Rev B 81(5). https://doi.org/10.1103/PhysRevB.81.054439
36. Yakata S et al (2010) Control of vortex chirality in regular polygonal nanomagnets using in-plane magnetic field. Appl Phys Lett 97(22):222503. http://aip.scitation.org/doi/10.1063/1.3521407
37. Vogel A et al (2012) Vortex dynamics in triangular-shaped confining potentials. J Appl Phys 112(6):063916
38. Krutyanskiy VL et al (2013) Second harmonic generation in magnetic nanoparticles with vortex magnetic state. Phys Rev B 88(9):094424. http://aip.scitation.org/doi/10.1063/1.4754418
39. Kida N et al (2005) Optical magnetoelectric effect in a submicron patterned magnet. Phys Rev Lett 94(7). https://doi.org/10.1103/PhysRevLett.99.197404
40. Agyei AK, Birman JL (1990) On the linear magnetoelectric effect. J Phys Condens Matter 2(13):3007–3020. https://doi.org/10.1088/0953-8984/2/13/010
41. Gartside JC et al (2016) A novel method for the injection and manipulation of magnetic charge states in nanostructures. Sci Rep 6(1):32864. https://doi.org/10.1038/srep32864

42. Jungwirth T et al (2018) The multiple directions of antiferromagnetic spintronics. Nat Phys 14(3):200–203. https://www.nature.com/articles/s41567-018-0063-6
43. Arregi JA et al (2018) Magnetization reversal and confinement effects across the metamagnetic phase transition in mesoscale FeRh structures. J Phys D Appl Phys 51(10). https://doi.org/10.1088/1361-6463/aaaa5a
44. Nisoli C (2018) Topology by design in magnetic nano-materials: artificial spin ice. In: Gupta S, Saxena A (eds) The role of topology in materials, vol 189. Springer, pp 85–112
45. Gartside JC et al (2018) Realization of ground state in artificial kagome spin ice via topological defect-driven magnetic writing. Nat Nanotechnol 13(1):53–58

Chapter 7
Optical Effects in Artificial Magneto-Toroidal Crystals

Both, linear magneto-optical effecfts [1–3], as introduced in Sect. 3.2.2, and the anomalous Hall effect [4] are based on the violation of time-reversal symmetry that manifests in a coupling to electromagnetic fields by energy-state shifts due to spin-orbit interaction. As a consequence, the material under consideration exhibits an anisotropic response to electromagnetic fields that relates to the direction of magnetic-field-induced or spontaneously-aligned magnetic moments, see Sect. 3.2.2. In contrast, the absence of point symmetries may allow for apparently similar macroscopic effects such as an optical birefringence or an optical activity. However, those effects are based on electric-field-induced or inherent crystal-field anisotropies with respect to atom- or ion-side symmetries in, e.g., a polar or chiral material.
A combination of these two symmetry constraints with different microscopic origin results in a variety of optical effects, namely the *magnetochiral effect* (MChE) that arises in a chiral medium with broken time-reversal symmetry and an *optical magnetoelectric effect* (OME) in a polar medium with broken time reversal symmetry. More precisely, the OME can be understood as a combined Pockels and Voigt effect, both inducing linear birefringence. As a consequence, the origin of the effect does not have to be an intrinsic material property such as in a polar magnetic matter, but can be induced by applied electric and magnetic fields [5]. Both OME and MChE have in common that they manifest themselves in non-reciprocal directional-dependent material responses, associated for example with the sign of the light's propagation vector in optical experiments [6–11]. The OME is typically probed as a change of optical absorption (α) upon a sign change of the light's wave-vector component k_i. In a polar magnetic material, for example, the sign of the effect depends on the sign of the triple product of the light's wave vector ($\vec{\mathrm{k}}$) and the sample's polarisation ($\vec{\mathrm{P}}$) and magnetisation ($\vec{\mathrm{M}}$) as $\frac{\Delta\alpha}{\alpha} \propto \vec{\mathrm{k}} \cdot (\vec{\mathrm{P}} \times \vec{\mathrm{M}})$. This difference is typically very small ($\lesssim 1\,\%$), because of the small magnitude of the microscopic source for the effect. The origin can be an optical interference between electric- and magnetic-dipole transitions, which usually have different resonance frequencies and oscillator strengths. Yet, if the two amplitudes corresponding to the resonances can be matched by the

J. Lehmann, *Toroidal Order in Magnetic Metamaterials*, Springer Theses,
https://doi.org/10.1007/978-3-030-85495-9_7

application of large magnetic (Zeeman effect) or electric (Pockels effect) fields, it can ultimately lead to a one-way transparency [12]. This eventually allows for optical applications as for instance magnetically controlled 'diodes', switches or filters for light beams.

The existence of such non-reciprocal directional anisotropies (NDA) have been either suggested and/or measured in single-phase linear magnetoelectric materials at selected $3d$–$3d$ transitions in CuB_2O_4 [5, 12–17], $LiCoPO_4$ [18], $CuFeO_2$ [19], $GaFeO_3$ [20, 21], Cr_2O_3 [22–24, 55, 56], Ni_3TeO_6 [25], $TlCuCl_3$ [26], $Ba_3NbFe_3Si_2O_{14}$ [27], $(Ca,Sr,Ba)_2Co(Si,Ge)_2O_7$ [28–31], as well as in artificial non-centrosymmetric magnets [32–40]. Magnetoelectric multiferroics that violate both space-inversion and time-reversal symmetry and, in addition, in which the two spontaneous orders that underlie the envisaged optical effect can be controlled by an external handle, see Sect. 2.1.2, are particularly interesting candidates. Further candidate systems are ferrotoroidic materials in which the effect may be used for an all-optical probing of the toroidisation or a spatially-resolved detection of domain pattern. Such a linear-optical approach to study ferrotoroidic order would provide a direct experimental access and may serve as a key phenomena for deriving possible applications.

As mentioned above, the magnitude of directional dichroism is typically very small in natural materials, which motivates the use of metamaterials composed of non-centrosymmetrically arranged magnetic building blocks, where resonances of both electric and magnetic multipoles can be tailored by geometry, size, material or arrangement of its constituents. Furthermore, enhancement through (particle-) plasmonic or lattice resonances, see Sect. 2.3 on page 39, can be advantageous for a large optical response. In mesoscopic polar and magnetic structures, the sub-micrometre periodicity implies resonances in the infrared spectral range while the required symmetry violations possibly originate solely from the magnetic order. This suggests coupled optical transitions of electric-, magnetic- and/or toroidal-dipole nature [9, 10]. Measurements of the OME in non-centrosymmetric magnetic nanostructures by Kida et al. [7, 32] have demonstrated a field-dependent NDA in the near-infrared spectral region. Here, the effect occurred in structural diffraction spots of the periodic array and exhibited a magnitude comparable with the one of the TMOKE signal. In their experiments, 'v'-shaped permalloy elements have been used that can be magnetised and comprise—as the authors call it—a 'virtual polarisation', which is in fact a characteristic direction that exists due to their asymmetric shape, see also Sect. 6.3. These arrays, however, neither allow for the emergence of a local toroidal moments, following the definition as given in Sect. 2.1.5 on page 17, nor exhibit a spontaneous toroidisation, either of which constitutes key properties of the ferrotoroidic state where the optical coefficients would relate to the macroscopic toroidal domain pattern. The measured effect relates to the outer vector product of a magnetisation and a polarisation [8] as shown above, which is symmetry-wise less constrained and does not comply with ferrotoroidicity.

Moreover, Udalov et al. [36] demonstrated a non-reciprocal optical effect in diffraction geometry on an artificial planar structure of equilateral sub-micrometre-sized cobalt and cobalt-iron triangles hosting magnetic vortex states. The effect has been

probed as changes in the intensity of a diffraction spot that lies in the scattering plane. However, the published results are inconclusive and allow for an ambivalent interpretation. Importantly, the geometry of the studied array intentionally prevents a magnetostatic coupling, while further it does not allow for a spontaneously emerging long-range-ordered state with a uniform toroidisation.
This chapter starts by introducing detection schemes to search for NDA (in this context referred to as *toroidal anisotropy* or more precise *toroidal dichroism*) in nanomagnetic toroidal arrays. A survey of two experimental attempts is given that deal with the surface-averaged and the spatially-resolved probing of a toroidal anisotropy. The averaged measurement, as discussed in the first part of the chapter, revealed optical signatures that appear to originate from the net toroidisation of the sample, while the second part describes the unsuccessful attempt to optically image a toroidal domain pattern. The chapter closes with a conclusion and a brief perspective of potential resonance-enhancement effects.

7.1 Integrated Detection of Toroidal Dichroism

In the experiments by Kida et al. [32] as well as Udalov et al. [36], as referred to in the introduction of this chapter, magneto-optical diffraction techniques have been applied for sensing NDA. These techniques provide principal advantages over measurement methods at a specular-reflected light beam only, see Sect. 3.2.2. Since each of the scattered beams correspond to a different direction of the final wave vector $\vec{k}$, a variety of $\vec{k}$-dependent phenomena can potentially be studied without the need for an exchange of source and detector, see Fig. 7.1 in Ref. [8].
For detecting directional dichroism in toroidally ordered metamaterials, an array of equilateral triangles has been used (see Fig. 4.1d) comprising permalloy building blocks of 400 nm edge length, 20 nm thickness, corner radii of 50 nm and separations of 100 nm. In exactly the same array, spontaneous domain formation has already been demonstrated (see Sect. 5.4). In addition, the ability to control the net toroidisation has been experimentally confirmed (see Sect. 6.3). A homogeneous magnetic field provides control over the sign of the net toroidisation $\vec{T}$ and is the key for repetitively probing an overall optical effect that corresponds to $\pm\vec{T}$. Note that, none of the macrospin-based arrays appear suitable for this type of experiment, since the generation of a suitable switching field in these has proven laborious, see Sect. 6.2.1. Therefore, the structurally non-centrosymmetric array of equilateral triangles as shown in Fig. 4.1d on page 83 constitutes an ideal system for measuring an optical anisotropy related to the toroidal state.
Shown in Fig. 7.1 is a magnetisation-reversal curve of the magnetic vortex array measured via LMOKE, see Sect. 3.2.2. The two observed steps in the hysteresis curve are known to exist in nanomagnetic elements that host (meta-)stable states between the two staurated regimes such as, here, magnetic vortex states: The energy penalty for the nucleation of a magnetic vortex state at the lower edge of the triangle is the cause for the non-zero coercive field. The extended plateau regions correspond

to the movement of the vortex core from the lower edge where it emerges towards the upper corner of the triangle where it finally annihilates. Note that both plateaus correspond to vortex states of uniform yet opposite vorticity. The respective offset is due to the non-zero applied magnetic field that deflects the vortex core from its centred equilibrium position and generates a net-magnetisation in each triangle, see the micromagnetic simulation (IV) in Fig. 7.1b. Besides this offset due to the deflection of the vortex state, no distinct asymmetry between the two toroidisation states occurs. The overall shape of the hysteresis curve merely makes it possible to distinguish the two saturated states from the two vortex states. Therefore, for sensing an optical signature of the sense of circulation of magnetic moments and, correspondingly, the net toroidisation of the sample, an advanced detection scheme is required.

A modification of the LMOKE setup (Fig. 3.7 on page 67) to sense magneto-optical effects in higher diffraction orders is shown in Fig. 3.9 on page 70. ΔDMOKE measurements as introduced in Sect. 3.2.2 have been performed with the sample placed in the aperture of a four-axis magnet, see Appendix B.2. The implementation of ΔDMOKE requires access to two diffraction peaks with either positive and negative diffracted wave-vector components k_i—members of so called *Friedel pairs* or *Bijvoet pairs* [42–44]. As shown in Fig. 7.2, a few diffraction spots are generated by the grating structure of the vortex-type toroidal array. Two of these diffraction spots with opposite k_y-value are scattered under a relatively small angle and can be monitored simultaneously using a balanced photodiode. Using a balanced detection of these two peaks allows for sensing the following:

(1) As introduced in Sect. 3.6, certain reflexes in the reciprocal space display changes when going from one toroidal domain state to the other. In an analogue way, any imbalance in the intensity of corresponding optical diffraction spots ($I(\pm k_i)$) may indicate an imbalance in the sample's net toroidisation ($\pm T$). Since such an effect would relate to a sign change in the final k vector, the interaction can be described in terms of an NDA.

(2) The array as such and each nanomagnetic triangle within the array structurally violates the mirror symmetry perpendicular to the y axis. For geometrical reasons, this m_y-mirror plane connects to the direction of the 'virtual polarisation' (a polar vector, see Sect. 2.1.1) [32] of each element. Following Friedel's law [42], generally known in x-ray crystallography, the reciprocal lattice of any structure will expose as being centrosymmetric if not measured resonantly [43, 45]. As a consequence, the structure itself can not cause an imbalance in the intensity of the two Friedel pairs $\pm k_y$. Though, a magneto-optical interaction with the sample indeed permits an imbalanced contribution to the peak intensities. Therefore, the two magnetised states (corresponding to axial vectors, see Sect. 2.1.1) along $\pm x$ (see Fig. 7.1I,IV) break m_y symmetry and yield different MOKE intensities in either of the Friedel-pair peaks. The balanced detection of the Friedel-pair intensity during a magnetic reversal along x, with incident light polarised along x and y, respectively, yields hysteresis curves equivalent with the ones shown in Fig. 7.1. The x-component of the magnetisation distri-

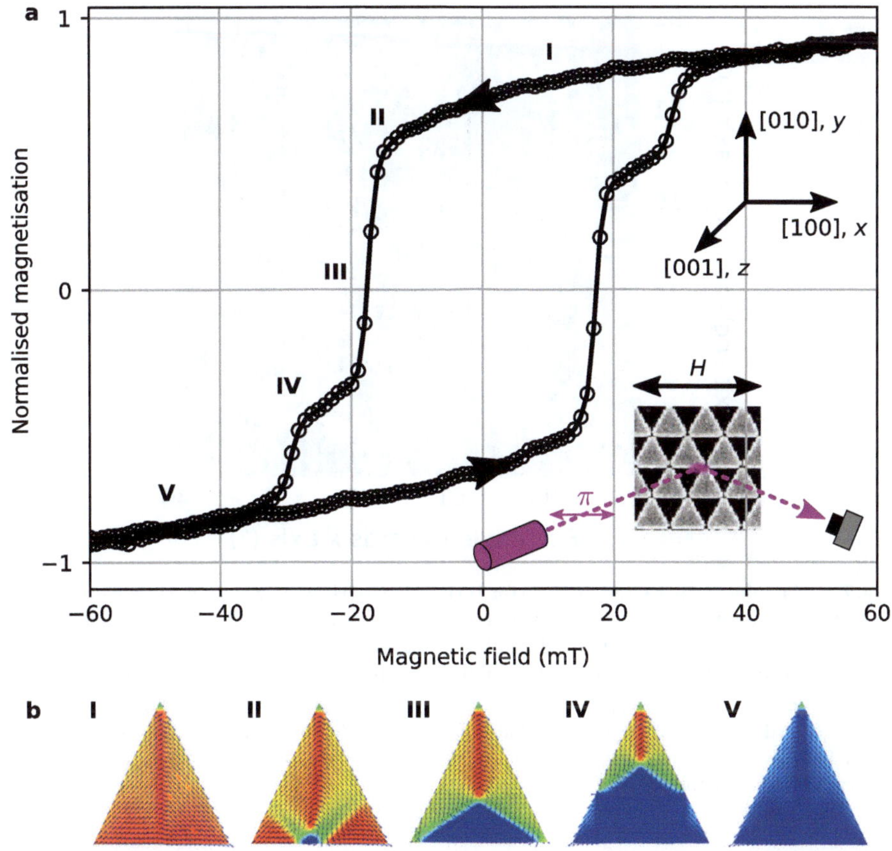

Fig. 7.1 Magnetisation curve of a vortex-type magneto-toroidal array. a Magnetisation curve measured on an array of coupled magnetic vortices as shown in the inset (Fig. 4.1d). LMOKE has been performed in the specular reflex with an external magnetic field along ±[100] generated by the four-axis magnet (see Appendix B.2). The laser light of 405 nm wavelength, 10 mW power and polarisation along [101] is incident under 45° ([10$\bar{1}$]) with respect to the sample normal and focussed to a spot diameter of 50 μm at the sample area to match the array size. **b** Micromagnetic simulation of field-induced changes in the internal magnetisation distribution of a nickel triangle, extracted from Ref. [41]. Shown is the nucleation (I–II) and annihilation (IV–V) of a magnetic vortex core in between saturated states with a magnetisation direction along [100] (I and V). The vortex state (III) defines the overall toroidisation of the array and manifests itself as plateau regions in the magnetisation curve

bution apparently dominates the magneto-optical interaction.[1] Furthermore, the y-mirror plane reverses the orientation of the out-of-plane magnetisation direction of the emergent magnetic vortex core, see Sect. 2.2.3. Thus, the balanced detection of the Friedel pair $\pm k_y$ can reveal the magnetisation direction of the vortex core with an up/down degree of freedom. An additional PMOKE signature,

[1] Measurements of a single diffraction spot are discussed in Appendix C.6.

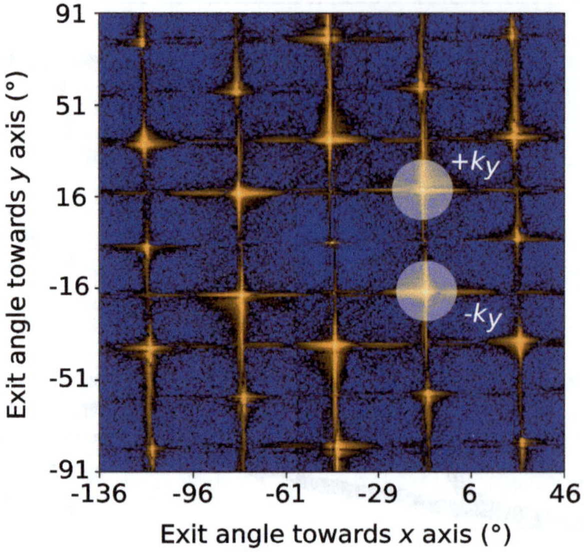

Fig. 7.2 Calculated diffraction pattern for DMOKE experiments. Shown is a calculated diffraction pattern resulting from Fourier transform of an AFM image of the magneto-toroidal array. The calculations imply the use of laser light with λ = 405 nm wavelength, incident under an angle of $\theta_{\mathrm{inc}} = 45°$ with respect to the sample normal in the xz-plane. The angles have been calculated as $\theta_{\mathrm{calc}} = \arccos(1 - \frac{G^2\lambda^2}{2}) + \theta_{\mathrm{inc}}$ from an Ewald construction, where G is a reciprocal lattice vector extracted via discrete Fourier transform of an AFM image of the structure and θ_{calc} are the calculated diffraction-spot angles. Highlighted are the two diffraction peaks under consideration for DMOKE and ΔDMOKE experiments

see Sect. 3.2.2, due to the out-of-plane oriented magnetic-vortex core is unlikely because it has not been seen in the DMOKE measurements performed in a single k_y reflex, see Appendix C.6 and, further, it would contribute symmetrically to both $\pm k_y$ diffraction spots.

(3) Any variation in the y periodicity (or the corresponding component in the magnetisation distribution m_y) would contribute to the intensity of the two diffraction spots. However, since both diffraction spots would experience the same variations, no net signal would show up in a balanced detection.

The intensity-balanced detection of diffraction spots ($\Delta I(k_y)$) using circularly polarised incident light, see Fig. 7.3, reveals distinct minima and maxima, located at the two plateau regions in the hysteresis curve shown in Fig. 7.1a. Accordingly, the signal reflects a stronger/weaker diffraction efficiency in either of the two diffraction spots $\pm k_y$. Since the studied magneto-toroidal array exhibits a net toroidisation at precisely these magnetic-field values, the measured signal seems to consists of an ordinary hysteresis curve superimposed with a signal that encodes the toroidisation as negative or positive peaks, respectively. As discussed above, however, this signal can originate from two phenomena related to the collectively formed magnetic vor-

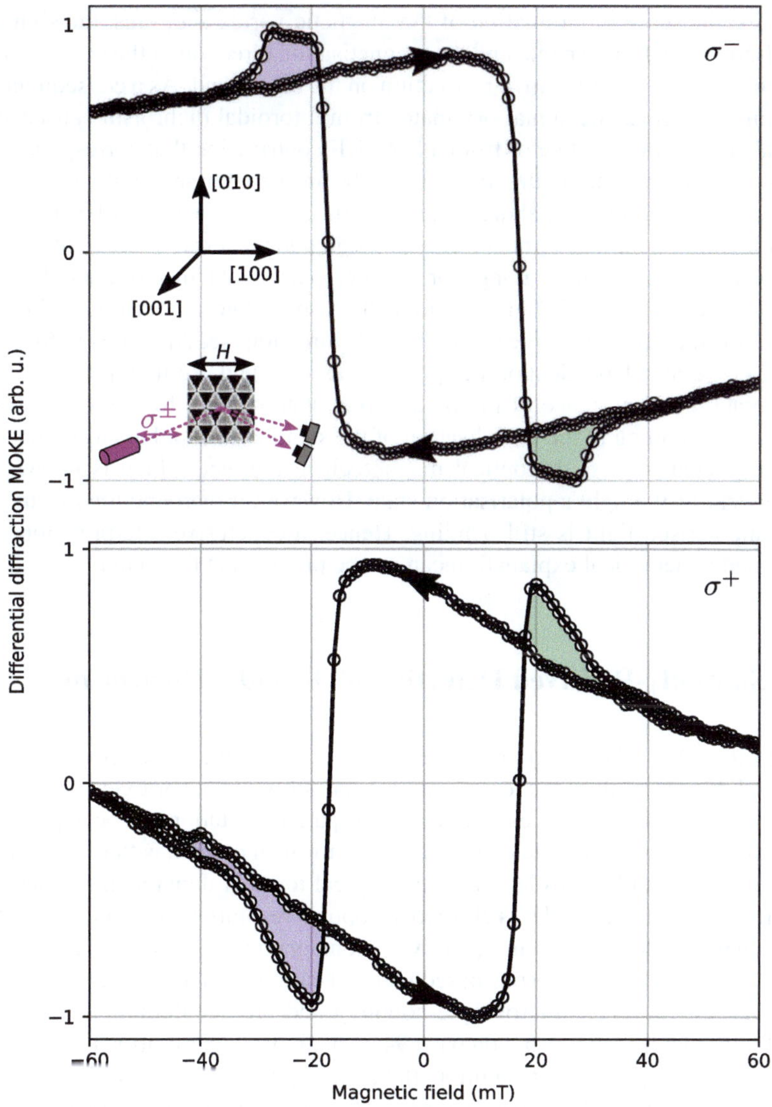

Fig. 7.3 Δ DMOKE measurements performed on a vortex-type magneto-toroidal array. The magnetic-field-dependent signal measured via an intensity comparison of the $+k_y$ and $-k_y$ diffraction spots, highlighted in Fig. 7.2. The measurement has been performed with a an external magnetic field along x and with laser light of 405-nm-wavelength, 10-mW-power and circular polarisation $\sigma^{\pm}$ incident under 45° ([10$\bar{1}$]) with respect to the sample normal. The single-domain states with toroidisation $-T$ (violet) or $+T$ (green) now appear as distinct minima and maxima

tex states: the sense of circulation of the magnetic vortex and, thus, the sign of the toroidisation, on the one hand, and the magnetisation direction of the vortex core that can point in negative or positive z direction on the other hand. As a consequence, the question of whether the signal originates from a toroidal dichroism caused by the toroidal single-domain state or from a PMOKE contribution that corresponds to the vortex-core-magnetisation direction can not be answered based on the results.

The nonreciprocal directional dichroism in diffraction geometry has been measured using circularly polarised light and changes sign upon reversal of the light's helicity. With linearly polarised light along y or in the xz-plane, I obtained ordinary hysteresis curves as shown in Fig. 7.1. In addition, the use of linear polarised light with a polarisation direction along the sample's [111] direction (see Appendix C.6) lead to a vanishing signal independent of the applied magnetic field. Both of these side results corroborate the importance of the polarisation state of light for the detection of a signal corresponding to the toroidisation of the sample. This polarisation sensitivity is not expected for e.g. the optical magnetoelectric effect, which is known to be independent of the light's polarisation state. However, a reproduction of this effect using unpolarised light is still pending. Hence, a conclusive interpretation of the results and a theoretical explanation can not be provided at this point.

7.2 Spatially-Resolved Detection of Toroidal Dichroism

In order to address the question of whether the origin of the signal discussed above, see Fig. 7.3), is a result of the toroidisation or possibly of the vortex-core magnetisation direction, a consecutive experiment with spatial resolution has been performed. The basic idea is an optical detection of a static toroidal domain pattern that bas been resolved beforehand by MFM. Since the typical toroidal domain size in structures made from magnetic triangles is close to the optical resolution limit, an optical detection appears challenging. Consequently, nanomagnetic systems with large toroidal domains provided by macrospin-based elements are preferred. These arrays exhibit an intrinsically stronger microscopic coupling and provide domain sizes of a few tens of micrometre, see Fig. 5.4 on page 94, well above the resolution limit of simple optical microscopy setups. In addition, the Ising-type arrays are inherently free from an out-of-plane magnetised vortex core such that a signal can not be caused by a PMOKE contribution from the vortex core.

The experimental setup is based on a modification of the setup for measuring an averaged signal as discussed above. Here, instead of detecting a signal in a diffraction spot with $k_y \neq 0$ upon incident circular-polarised light, the propagation vector of light from a particular diffraction spot is used for illuminating the sample with either a collimated or a focussed light beam. Conversely, the former illumination path is now used for imaging with a 10× microscope objective and a selectable polarisation filtering. For live imaging, a black/white CMOS camera (Imaging Source DMK 22BUC03) images the array, while a liquid-nitrogen-cooled CCD camera (Photometrics A2822) is used for a more sensitive and less noisy detection. The setup, thus, uses a variance of dark-field microscopy in which light can be detected only if a certain accordingly aligned sample is mounted.

The experiment has been performed on a toroidal square array exhibiting large toroidal domains, see Fig. 5.4a. A typical reflection microscopy setting has been used to localise the desired array on the silicon chip and to position and focus the sample. Then, laser light of 650 nm wavelength has been used in normal incidence along the illumination beam path to identify the angle, direction and accessibility of diffraction peaks. Subsequently, a second laser of identical wavelength, yet with variable polarisation (linear and circular), has been guided back antiparallel to the chosen diffraction spot. Imaging has been performed using only this second laser light for illumination in a dark-field configuration to possibly reveal signatures of the distribution of the sample's toroidisation and, accordingly, the sample's domain structure. For an improved signal-to-noise ratio, differential imaging has been implemented via the subtraction of images. This enhances possible differences in subsequently captured frames with different input-polarisation settings.

However, besides imaging of the nanomagnetic array as such, no contrast between different toroidal domain states was achieved. The reason for this may be, on the one hand, an absence of magnetic vortex cores, assuming that the vortex-core magnetisation is the origin for the signal that has been detected in the integrated measurement, see the section above. On the other hand, the searched signature might be a very small optical effect that requires, even when measured with an averaging technique as discussed in the section above, multiple superimposed measurements. Furthermore, the averaging measurement of NDA was performed by a simultaneous detection of two diffraction spots, while in the imaging setup, a single k-vector has been accessed.

7.3 Summary and Perspective

The ferrotoroidic state that is odd under space-inversion and time-reversal inherently allows for exotic optical responses such as a nonreciprocal directional dichroism [7, 8], which is the dependence of the absorption of light on the direction of propagation, eventually resulting in an 'optical diode' [12]. Due to the ferroic nature of considered crystals, the optical effect correlates directly with the direction of the order parameter and, therefore, depends on the macroscopic domain structure. Here, ferrotoroidic arrays of nanomagnets provide a new route for accessing such non-reciprocal optical phenomena. Those effects, which are based on the change of direction of the light propagation k, can be measured background-free and with beneficial degrees of freedom using a simple diffraction geometry. Initially, magnetisation curves have been measured using LMOKE on an array that consists of triangular nanomagnets. The measurements provide a means to identify magnetic-field values that are required to induce a magnetic vortex state in each triangle. Then, an intensity-balanced magnetic-field-dependent measurement of two diffraction spots with opposite k_y-value has been performed with circular polarised light. The signal corresponds to the reversal of light propagation along the y axis of the crystal. The measurement revealed a non-reciprocal signal that showed minima or maxima at magnetic-field values where a $-T$ or $+T$ net toroidisation emerges, respectively.

Two explanations for the observed effect have been proposed; the toroidisation that relates to the handedness of the vortex state on the one hand and the out-of-plane magnetisation of the vortex core on the other hand. This uncertainty about the microscopic origin of the effect provides a scope for further experiments. The signal's origin needs to be studied during magnetisation reversal for instance by micromagnetic simulations in combination with a Fourier-transform-based analysis of the y-magnetisation component. In addition, the application of a small bias magnetic field in a defined out-of-plane direction would limit the signal to originate from a specific vortex-core-magnetisation direction. In an analogue way, a diffraction experiment that utilises triangle-shaped nanomagnetic elements that avoid the existence of a vortex core by having a hole in their centres may give an indication for the signal's origin.

An experiment for the spatially-resolved detection of toroidal domain patterns has been conceived and tested, without being able to optically image a toroidal domain pattern at this point. A final goal for further experiments would remain to achieve spatial resolution and to access a sample's toroidisation by linear-optical methods. This may require an enhancement of the underlying effects. To address this point, flexibility in designing the sub-micrometre-sized building blocks and the related periodicity of the arrays would be beneficial. As introduced in Sect. 2.3, metallic structures can host plasmonic resonances and lattice resonances, which are potentially accompanied by a field enhancement and an increased sensitivity to the studied effect particularly by spectrally addressing certain resonance features [9, 46]. In this context, the building blocks of the structure can be considered as nano-antennas or connected LC circuits in which the magnetisation enhances or weakens the electromagnetic coupling between neighbouring nanomagnets and may form collective modes [47–49].

Linear-optical transmission and reflection experiments,[2] however, revealed no spectral features for the different arrays that significantly deviate from the bare silicon substrate. This may be due to the low optical conductivity [50] of permalloy [51, 52] and thus, high damping factors for any type of plasmonic resonance. In addition, silicon as substrate material exhibits a high refractive index that significantly red-shifts, damps and broadens electromagnetic resonances [53, 54]. Nevertheless, indications for plasmonic resonances have been found during optical damage-threshold tests using laser pulses. Here, a 3-nm-thick artificial spin-ice structure has been treated with intense laser pulses and showed a selective interaction with just a particularly-aligned sublattice of the nanomagnetic structure, see Appendix C.7. This finding provides a motivation for ensuing spectroscopic studies of thin magneto-toroidal arrays

[2] Linear-optical spectroscopy has been performed from the ultraviolet to the near infrared spectral region using a JASCO MSV-370 microspectrometer in our laboratories. Furthermore, the spectral region from the mid infrared to the far infrared has been probed using a Bruker 80V FTIR in the laboratories of Prof. Dr. Leonardo Degiorgi at ETH Zurich.

deposited on low-refractive-index substrates such as glass. Corresponding samples may allow for a resonantly enhanced interaction with the array and an associated increase in the toroidal dichroism.

References

1. Zvezdin AK, Kotov VA (1997) Modern magnetooptics and magnetooptical materials. Institute of Physics Publication
2. Kronmueller H, Parkin S (eds) (2007) Handbook of magnetism and advanced magnetic materials, 1st edn. Wiley. http://dx.doi.org/10.1002/9780470022184
3. Hubert A, Schaefer R (2009) Magnetic domains: the analysis of magnetic microstructures. Springer
4. Nagaosa N et al (2010) Anomalous Hall effect. Rev Mod Phys 82(2):1539–1592. https://doi.org/10.1103/RevModPhys.82.1539
5. Saito M et al (2008) Magnetically controllable CuB_2O_4 phase retarder. Appl Phys Exp 1. https://doi.org/10.1143/APEX.1.121302
6. Baranova NB, Bogdanov YV, Zel'dovich BY (1977) New electro-optical and magneto-optical effects in liquids. Soviet Phys Uspekhi 20(10):870–877. https://mr.crossref.org/iPage?doi=10.10702FPU1977v020n10ABEH005470
7. Arima T (2008) Magneto-electric optics in non-centrosymmetric ferromagnets. J Phys Condens Matter 20(43):434211. https://doi.org/10.1103/PhysRevB.70.064426
8. Szaller D, Bordacs S, Kezsmarki I (2013) Symmetry conditions for nonreciprocal light propagation in magnetic crystals. Phys Rev B 87(1):014421. https://doi.org/10.1103/PhysRevB.87.014421
9. Tomita S et al (2018) Metamaterials with magnetism and chirality. J Phys D Appl Phys 51(8):083001. https://doi.org/10.1088/1361-6463/aa9ecb
10. Cheong S-W et al (2018) Broken symmetries, non-reciprocity, and multiferroicity. Npj Quant Mater 3(1):19. http://dx.doi.org/10.1038/nmat1804
11. Tokura Y, Nagaosa N (2018) Nonreciprocal responses from non-centrosymmetric quantum materials. Nat Commun 9(1):3740. https://doi.org/10.1088/0034-4885/77/7/076501
12. Toyoda S et al (2015) One-way transparency of light in multiferroic CuB_2O_4. Phys Rev Lett 115(26):267207. https://doi.org/10.1103/PhysRevLett.115.267207
13. Saito M, Taniguchi K, Arima T-H (2008) Gigantic optical magnetoelectric effect in CuB_2O_4. J Phys Soc Jpn 77(1):013705
14. Saito M et al (2008) Magnetic control of crystal chirality and the existence of a large magneto-optical dichroism effect in CuB_2O_4. Phys Rev Lett 101(11). https://doi.org/10.1143/JPSJ.77.013705
15. Toyoda S, Abe N, Arima T (2016) Gigantic directional asymmetry of luminescence in multiferroic CuB_2O_4. Phys Rev B 93(20):201109. https://doi.org/10.1103/PhysRevB.93.201109
16. Toyoda S, Abe N, Arima T (2019) Nonreciprocal refraction of light in a magnetoelectric material. Phys Rev Lett 123(7):077401. https://doi.org/10.1103/PhysRevLett.123.077401
17. Toyoda S et al (2020) Nonreciprocal second harmonic generation in a magnetoelectric material. arXiv 2006.01728. http://dx.doi.org/10.1103/PhysRevLett.115.267207
18. Kocsis V et al (2018) Identification of antiferromagnetic domains via the optical magnetoelectric effect. Phys Rev Lett 121(5):057601. https://doi.org/10.1103/PhysRevLett.121.057601
19. Kibayashi S et al (2014) Magnetochiral dichroism resonant with electromagnons in a helimagnet. Nat Commun 5(1):4583. https://doi.org/10.1038/ncomms5583
20. Jung JH et al (2004) Optical magnetoelectric effect in the polar $GaFeO_3$ ferrimagnet. Phys Rev Lett 93(3):037403. https://doi.org/10.1103/PhysRevLett.93.037403
21. Igarashi J-I, Nagao T (2009) Analysis of optical magnetoelectric effect in $GaFeO_3$. Phys Rev B 80(5):054418. https://doi.org/10.1103/PhysRevB.80.054418

22. Pisarev RV, Krichevtsov BB, Pavlov VV (1991) Optical study of the antiferromagneticparamagnetic phase transition in chromium oxide Cr_2O_3. Phase Trans 37(1):63–72. https://www.tandfonline.com/doi/abs/10.1080/01411599108203448
23. Krichevtsov BB et al (1993) Spontaneous non-reciprocal reflection of light from antiferromagnetic Cr_2O_3. J Phys Condens Matter 5(44):8233–8244. https://iopscience.iop.org/article/10.1088/0953-8984/5/44/014
24. Goulon J et al (2002) X-Ray magnetochiral dichroism: a new spectroscopic probe of parity nonconserving magnetic solids. Phys Rev Lett 88(23):237401. https://doi.org/10.1103/PhysRevLett.88.237401
25. Yokosuk MO et al (2020) Nonreciprocal directional dichroism of a chiral magnet in the visible range. npj Quan Mater 5(1):20. http://dx.doi.org/10.1038/s41535-020-0224-6
26. Kimura S, Matsumoto M, Tanaka H (2020) Electrical switching of the nonreciprocal directional microwave response in a triplon bose-Einstein condensate. Phys Rev Lett 124(21):217401. https://doi.org/10.1103/PhysRevLett.124.217401
27. Narita H et al (2016) Observation of nonreciprocal directional dichroism via electromagnon resonance in a chiral-lattice helimagnet $Ba_3NbFe_3Si_2O_{14}$. Physical Review B 94(9):094433. https://doi.org/10.1103/PhysRevB.94.094433
28. Kezsmarki I et al (2011) Enhanced directional dichroism of Terahertz light in resonance with magnetic excitations of the multiferroic $Ba_2CoGe_2O_7$ oxide compound. Phys Rev Lett 106(5):057403. https://doi.org/10.1103/PhysRevLett.106.057403
29. Bordacs S et al (2012) Chirality of matter shows up via spin excitations. Nat Phys 8(10):734–738. https://www.nature.com/articles/nphys2387
30. Kezsmarki I et al (2014) One-way transparency of four-coloured spin-wave excitations in multiferroic materials. Nat Commun 5(1):3203. https://www.nature.com/articles/nphys2387
31. Viirok J et al (2019) Directional dichroism in the paramagnetic state of multiferroics: a case study of infrared light absorption in $Sr_2CoSi_2O_7$ at high temperatures. Phys Rev B 99(1):014410. https://doi.org/10.1103/PhysRevB.99.014410
32. Kida N et al (2005) Optical magnetoelectric effect in a submicron patterned magnet. Phys Rev Lett 94(7):077205. https://doi.org/10.1103/PhysRevLett.94.077205
33. Kida N et al (2006) Enhanced optical magnetoelectric effect in a patterned polar ferrimagnet. Phys Rev Lett 96(16):167202. https://doi.org/10.1103/PhysRevLett.96.167202
34. Kida N et al (2007) Optical magnetoelectric effect of patterned oxide superlattices with ferromagnetic interfaces. Phys Rev Lett 99(19):197404. https://doi.org/10.1103/PhysRevLett.99.197404
35. Levy M, Jalali AA, Huang X (2009) Magnetophotonic crystals: nonreciprocity, birefringence and confinement. J Mater Sci Mater Electron 20(S1):43–47
36. Udalov OG et al (2012) Nonreciprocal light diffraction by a lattice of magnetic vortices. Phys Rev B 86(9):094416. https://doi.org/10.1103/PhysRevB.86.094416
37. Figotin A, Vitebsky I (2001) Nonreciprocal magnetic photonic crystals. Phys Rev E 63(6):066609. https://doi.org/10.1103/PhysRevE.63.066609
38. Figotin A, Vitebskiy I (2003) Electromagnetic unidirectionality in magnetic photonic crystals. Phys Rev B 67(16):165210. https://doi.org/10.1103/PhysRevB.67.165210
39. Figotin A, Vitebskiy I (2006) Electromagnetic unidirectionality and frozen modes in magnetic photonic crystals. J Magn Magn Mater 300(1):117–121. https://doi.org/10.1016/j.jmmm.2005.10.046
40. Eslami S et al (2014) Chiral nanomagnets. ACS Photon 1(11):1231–1236. https://doi.org/10.1021/ph500305z
41. Jaafar M et al (2008) Field induced vortex dynamics in magnetic Ni nanotriangles. Nanotechnology 19(28). https://doi.org/10.1088/0957-4484/19/28/285717
42. Friedel G (1913) Sur les symétries cristallines que peut révéler la diffraction des rayons Röntgen. Comptes Rendus 157:1533–1536
43. Bijvoet JM, Peerdeman AF, van Bommel AJ (1951) Determination of the absolute configuration of optically active compounds by means of X-rays. Nature 168(4268):271–272. https://doi.org/10.1038/168271a0

44. http://skuld.bmsc.washington.edu/scatter/AS_Friedel.html. (Online content, Accessed on 13 Mar 2020)
45. Gorfman S et al (2016) Simultaneous resonant x-ray diffraction measurement of polarization inversion and lattice strain in polycrystalline ferroelectrics. Sci Rep 6(1):20829. https://doi.org/10.1038/srep20829
46. Shaltout AM, Shalaev VM, Brongersma ML (2019) Spatiotemporal light control with active metasurfaces. Science 364(6441):1. https://doi.org/10.1126/science.aat3100
47. Linden S et al (2006) Photonic metamaterials: magnetism at optical frequencies. IEEE J Select Top Quant Electron 12(6):1097–1105. https://doi.org/10.1109/JSTQE.2006.880600
48. Liu N, Giessen H (2010) Coupling effects in optical metamaterials. Angewandte Chem Int Edn 49(51):9838–9852. https://doi.org/10.1002/anie.200906211
49. Gentile M et al (2011) Investigation of the nonlinear optical properties of metamaterials by second harmonic generation. Appl Phys B 105(1):149–162. https://doi.org/10.1007/s00340-011-4766-y
50. Cortie MB, Arnold MD, Keast VJ (2020) The quest for zero loss: unconventional materials for plasmonics. Adv Mater 32(18):1904532. https://onlinelibrary.wiley.com/doi/10.1002/adma.201904532
51. Blaber MG, Arnold MD, Ford MJ (2010) A review of the optical properties of alloys and intermetallics for plasmonics. J Phys Condense Matter 22(14):143201. https://doi.org/10.1088/0953-8984/22/14/143201
52. Rycenga M et al (2011) Controlling the synthesis and assembly of silver nanostructures for plasmonic applications. Chem Rev 111(6):3669–3712. https://doi.org/10.1021/cr100275d
53. Auguie B et al (2010) Diffractive arrays of gold nanoparticles near an interface: critical role of the substrate. Phys Rev B 82(15). https://doi.org/10.1103/PhysRevB.82.155447
54. Sadri-Moshkenani P et al (2020) Effect of magnesium oxide adhesion layer on resonance behavior of plasmonic nanostructures. Appl Phys Lett 116(24):241601. http://aip.scitation.org/doi/10.1063/5.0008665
55. Hornreich RM (1968) Gyrotropic birefringence-phenomenological theory. J Appl Phys 39(2):432–434. http://aip.scitation.org/doi/10.1063/1.2163466
56. Hornreich RM, Shtrikman S, Theory of gyrotropic birefringence. Phys Rev 171(3):1065–1074. https://journals.aps.org/pr/abstract/10.1103/PhysRev.171.1065

Chapter 8
Conclusion

The goal of this work was to establish experimental pathways for investigating ferrotoroidicity in appropriately engineered two-dimensional arrays composed of nanomagnetic building blocks. The study was based on scaling up both the individual spins and the underlying interaction mechanisms to mesoscopic length scales. This approach provides a means to arrange magnetic moments on demand and to substitute magnetic-exchange interactions with the geometrically determined and therefore more accessible magnetostatic interaction. As a consequence, the approach overcame limitations and experimental challenges of conventional atomic systems such as the elusiveness to detect and manipulate ferrotoroidic order. To achieve my goal, I designed arrangements of Ising-like macrospins and magnetic-vortex elements with a spontaneous toroidisation. These artificial nanomagnetic crystals were the basis of this work. After the refinement of design parameters, I worked out the key properties of a ferroic state: the spontaneous formation of domains and their manipulation via a conjugate field.

To improve existing functionalities and to develop novel applications of ferroics in general, inherent mechanisms that cause the emergence and determine the spatial distribution of domains have to be disclosed. I revealed that the size and morphology of ferrotoroidic domains in my systems are determined by the interplay of two complementary intrinsic interaction parameters. By tuning these parameters throughout a series of nanomagnetic arrays, I achieved to control both the domain size as well as the preferential spatial alignment of domain walls. In particular, I found that the domain walls in ferrotoroidic systems show similarities to Bloch walls in ferromagnets and, as such, allow for a substructure within the domain wall. This substructure manifests itself in the orientational degree of freedom of magnetic moments that enclose a toroidal domain. I succeeded in tuning between left-handed, right-handed and mixed/alternating types of domain walls by changing the ratio between the two interaction parameters. Especially the third domain-wall type merits attention due to its association with local magnetic charges, a type of topological defect, that

J. Lehmann, *Toroidal Order in Magnetic Metamaterials*, Springer Theses,
https://doi.org/10.1007/978-3-030-85495-9_8

occurs at the positions where the handedness of the domain wall reverses. I explained the observations with a simple model based on the energy and degeneracy of local ensembles of magnetic moments in the arrays. A result of this model is the formation of emergent magnetic subunits, namely quadrupoles and octupoles, that interact to promote and establish the spontaneous long-range order. My findings provide new general insight into the formation of domains in magnetically ordered systems that are not governed by the demagnetising-field energy. This accounts for the toroidal systems as well as for antiferromagnets or other types of magnetically compensated order that are envisaged to form the basis of an advanced spintronic memory and computing technology. Remarkably, my results include the first direct observation of the peculiar nature of the domain walls in ferrotoroidic crystals and, with this, extends the knowledge about ferrotoroidicity as such.

A next step in the study of toroidal domain properties may be to go towards thermally active systems in combination with a suitable observation method such as photoemission electron microscopy. This would allow one to observe the dynamics of toroidal domains such as the relaxation of excited states or the response upon the application of external fields.

In addition to the investigation of as-grown domain patterns and domain-wall structures, different schemes for manipulating toroidal order have been studied. The ability to manipulate a material-inherent property such as the toroidisation with an external handle lifts the potential for application from passive use, as for instance in sensors, towards active use as for instance in actuators or memory devices. However, poling of a toroidal order faces the challenge of generating a conjugate field, e.g. a magnetic-vortex field at the level of a unit cell. The generation of such a field appears elusive considering conventional atomic crystals. Yet, I demonstrated that it is tangible at mesoscopic length scales. I developed a switching scheme that, instead of an application of external fields, utilises magnetic fields that originate from the magnetostatic interaction within the strongly coupled nanomagnetic array. The activation of these local magnetic fields has been accomplished by scanning the array with a magnetised tip of a magnetic force microscope. The tip sequentially catalyses reorientation processes of individual magnets, while the trajectory of the tip determines the sign of a locally generated effective toroidal field that acts within the scanned area. Effective magnetic vortex fields of either sign have been generated and used for writing toroidal domains of different orientation into an artificial nanomagnetic array. I evaluated the poling mechanism with micromagnetic calculations and confirmed the vortex-like nature of the poling field. This unconventional mechanism for poling ferroic order may inspire similar approaches to generate otherwise elusive field configurations. Yet, if desired, the likelihood of transferring the conceived poling scheme back to atomic crystals is uncertain. Other switching schemes utilising macroscopic external poling-field configurations, that may represent conjugate toroidal fields as well, turned out not to be applicable at mesoscopic length scales. Nevertheless, these methods could find their application in conventional systems at the atomic scale.

The particular magnetic structure of the ferrotoroidic state implies an inherent optical magnetoelectric effect. This phenomenon results in a light-matter interaction with the toroidal system that depends on the direction of light propagation. By accessing different propagation directions of light provided via a diffraction process from the sample surface, I was able to identify an optical signature related to the toroidal state. These preliminary results motivate further studies to explore and confirm the microscopic origin of this phenomenon. Moreover, resonances in the interaction with the structure have to be exploited for ultimately achieving 'optical-diode'-like system characteristics.

On a fundamental level, the effective scaling up of spins and interactions may limit the generalisability and adaptability of the gained results. However, my approach provided a deep insight into properties of the ferrotoroidic state that would be virtually impossible to obtain with conventional methods applied to atomic crystals. Moreover, the investigated nanomagnetic systems do not have to be seen as just model systems, but they are an actual class of matter with their own practical and didactical advantages and disadvantages. In this context, it is reasonable to identify nanoscopic models that represent other types of ferroic or, in general, ordered states of matter to benefit from the unparalleled accessibility at the mesoscale. Importantly, the road that I have chosen here in conjunction with simulations, calculations and further experiments from solid-state chemists, materials scientists and condensed-matter physicists will help to shift the boundary of knowledge about novel ordered states of matter.

Appendix A
Development of Teaching Concepts

During my work I strived to gather many different teaching experiences. Besides supervising laboratory courses in non-linear and crystal optics, teaching exercise classes in crystallography and taking over several lectures for materials science students, I developed novel methodical approaches for teaching. Two of these contributions will be presented in the following. First, a device was conceived and built to demonstrate crystal diffraction and second, a sound-card-based circuit for data acquisition was set up.

A.1 Teaching Crystal Diffraction with Spatial Light Modulators

The reciprocal space, which is of tremendous importance in condensed-matter physics, is a challenging concept for undergraduate students in physics, chemistry or materials science. Students mostly struggle in thinking in terms of wave vectors and in intuitively 'feeling' the consequence of real-space periodicities and symmetries on the Fourier space. The reciprocal space is a central element in the description of crystal diffraction. Here, a collimated beam of either (x-ray) photons, electrons or neutrons interacts with a crystalline material. The elastic scattering process of the incoming radiation with a periodic ensemble of atoms in the material leads to the well known phenomena of diffraction. Students get a thorough theoretical introduction into the field, while their experimental access to the topic is very limited. If no 'hands-on' knowledge is conveyed in laboratory courses that address Laue or Debye-Scherrer diffraction, the direct experience is to a large extend restricted to computational or optical experiments serving as a substitution.
In those optical simulations both the radiation wavelength and the sample structures are upscaled from a few Angstrom to a few micrometre. Therefore a conventional laser pointer can be used as a light source substituting, for example, the x-ray tube, while a transparent slide patterned with printed ink dots can be used as a substitute for

J. Lehmann, *Toroidal Order in Magnetic Metamaterials*, Springer Theses, https://doi.org/10.1007/978-3-030-85495-9

the atomic scatterers in a crystalline solid. The patterns used in those experiments typically consist of periodic arrangements of dots representing the basic crystal lattices. Such an optical representation of a crystal-diffraction experiment provides a very simplified yet scientifically valuable basis for a live demonstration of the technique. To overcome limitations of this approach as being inflexible, non-interactive and restricted to the patterns that have been printed previously to the lecture, I resorted to a more advanced technical realisation of the underlying idea. My approach is a substitution of ink-dot-patterned transparencies with a programmable scattering pattern. This can be done by using spatial light modulators (SLMs)—typically thumbnail-sized microdisplays with pixel periodicities of sub-10 μm. An SLM can be controlled with a computer as an ordinary monitor. It allows for representing any two-dimensional scattering landscape by a corresponding spatial distribution of grey values within the displayed image. Accordingly, physical diffraction pattern can be generated, in real time, of any two-dimensional structure that can be displayed on a computer screen, including static images, animations or even interactive structural models from a crystallographic visualisation software such as VESTA [1, 2].
A detailed description of the concept, its realisation and suggestions for a didactic usage can be found in our publication in the JOURNAL OF APPLIED CRYSTALLOGRAPHY [3]. In order to reach more teachers or interested students with the concept, a cost-effective realisation has been strived for—money is a significant factor for school- or lab-course inventories. Therefore, the used SLMs have been extracted from commercially available video projectors. The working principle of the modified projector is explained and technical advice for its choice and successful modification is given. Furthermore, a PYTHON-based open-source program code has been developed in collaboration with my colleague Dr. Christian Tzschaschel and is provided together with instructions and explanations of its principle. The Program features the design and interactive manipulation of two-dimensional crystal structures that allow for the demonstration of concepts such as reciprocal space, structure factors, selection rules, symmetry and symmetry violation, as well as structural disorder.
The approach has already proved helpful in teaching crystal diffraction to undergraduate students in materials science within my teaching duties. Furthermore, the setup has been displayed as part of an exhibition at an international crystallography meeting.

A.2 Outline for a Data-Acquisition Lab Course at ETH Zurich

The automated and computer-based acquisition of data is a central element in all experiment-based natural sciences. I envisaged a lab course that focusses on the question of 'how to communicate an experimental setup and how to automatically acquire and possibly post-process data?'. The idea of a low-cost data acquisition (DAQ) system may be addressed within an electrical-engineering-centred lab course dealing with, for example, complex electronic systems such as loudspeakers and electronic filter networks. The computer-based DAQ system should be realised without the need for sophisticated and expensive hardware, while the approach should motivate the do-it-yourself mentality of the experimentalist.

General-Purpose Application of Computer Hardware

The idea to extend the range of use for dedicated computer hardware towards general-purpose applications is not new. Dedicated computer hardware such as graphics cards are known to serve not only for rendering and texturing frames for computer vision, but they can be used for performing specific tasks that benefit from the massively parallel computing power of a large amount of stream processors. This general-purpose computing on graphic cards finds applications in modern super computers for big data analysis in science and for the weather forecast as well as in the calculus of hash-functions for blockchain-based digital technologies such as crypto currencies. For a simple DAQ system, the usage of a computer sound card for general-purpose applications has been conceived. The beneficial properties of a computer sound card lays in its basic operation principle to serve as an electrical input-output system. Utilised may be the sound card's ability to synthesize arbitrary voltage wave forms in a broad frequency range from typically sub-Hz to $>100\,\text{kHz}$ in combination with simultaneous recording capabilities, both—depending on the particular ADC-DAC chip—with an amplitude resolution of commonly 24 bit. This basic input-output functionality with typically multiple audio (data) channels can be combined with the flexibility of programming the desired output signals and post-processing the resulting input signals of the card using software. This approach offers a vast playground for scientific use, in particular, for teaching technically affine students in a lab-course environment.

Impedance Spectroscopy Using a Sound Card

Impedance spectroscopy is broached by many different fields of science, since it governs the frequency-dependent resistance R and reactance χ for a general description of wave-associated phenomena. The value of interest is the total complex impedance $Z = R + i\chi$ of a system at interfaces or of the medium in which the wave propagates. Typical applications range from optical phenomena ($Z \propto$ the permeability and permittivity) over electronics ($Z \propto$ voltage and current), acoustics ($Z \propto$ air pressure and particle velocity), mechanics ($Z \propto$ mech. force and velocity), to gravitational-wave physics ($Z \propto$ speed of light and gravitational constant) and many more.
Here, an external two-channel USB sound card (*Focusrite Scarlett 2i2*) was used for measuring the impedance of simple electronic filter networks as well as of a loudspeaker chassis. The sound card was programmed using the computer language PYTHON with the libraries *sounddevice* for accessing the sound processor, *numpy* and *scipy* for signal generation, processing and evaluation as well as *matplotlib* for the visual representation of the results.

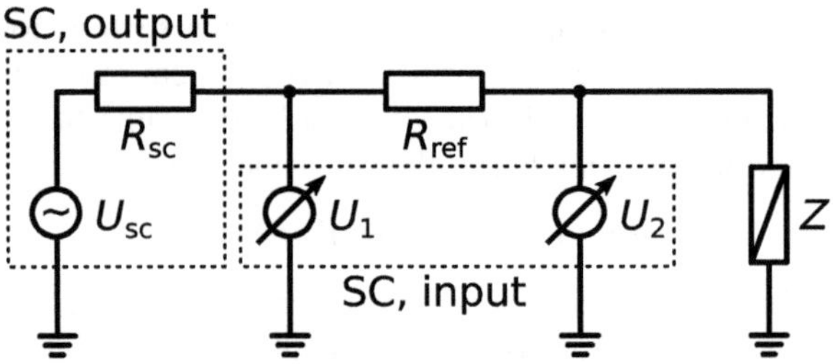

Fig. A.1 Circuit diagram for electrical impedance measurements. The output channel of the sound card (SC) with internal resistance R_{sc} drives a series circuit of a reference resistor R_{ref} and the component with unknown impedance Z. The synchronous measurement of two voltages by the sound-card's input channels allows an extraction of Z

The electrical realisation of the experiment is depicted in Fig. A.1. The single-channel voltage output of the sound card U_{sc} delivers an unknown current that depends on its current-gain function, internal resistance and reactance Z_{sc}. To yield a constant-current output from the constant-voltage output of the sound card, a reference precision resistor R_{ref} of 1 kΩ is connected in series to the output channel of the sound-card. The value of this resistor is chosen such that it dominates the total resistance R_{tot} and, accordingly, the total current I_{tot} flowing through the circuit. The unknown impedance Z is connected in series to this resistor. Two voltages are measured simultaneously using the (stereo) input-channels of the sound card (see Fig. A.1). This allows for an internal referencing of the voltages to extract a value for Z independent on the sound cards spectral characteristics as for instance frequency-dependent driving-voltage variations (so-called sound signatures). The channel U_1 accesses the voltage drop over the whole (external) circuit, while the channel U_2 accesses the voltage drop over the unknown impedance only. These two voltages are determined by

$$U_1 = R_{tot} \cdot I_{tot} = (R_{ref} + Z) \cdot I_{tot} \quad \text{and} \quad U_2 = Z \cdot I_{tot}. \tag{A.1}$$

Accordingly, by combining both equations, the unknown impedance Z can be determined as

$$Z = R_{ref} \cdot \frac{U_2}{U_1 - U_2}. \tag{A.2}$$

The experimental setup is shown in Fig. A.2. It can now be used for exciting an electrical system with a desired signal and for measuring its frequency-dependent response as voltage drops in two channels. For accessing spectral characteristics of the examined system, different excitation schemes can be used, for example single-frequency excitations for a high accuracy at a particular frequency, Dirac pulses or pulse trains, different types of noise, linear or logarithmic chirps or even arbitrary broad-band signals such as music for accessing a complete spectral map. Here, logarithmic chirps are generated representatively for probing the examined system at a broad spectral window between 10 Hz and 20 kHz. The measured voltages are analysed via a Fourier

Fig. A.2 Simple experimental setup for impedance-spectroscopy measurements. The USB sound card (red) is connected to a small measurement box containing the reference resistor, an internal calibration resistor and a terminal for connecting the electrical component of interest

transform in order to retrieve spectral dependencies of the investigated components. For simplicity, the phase has not been extracted as a separate property. Measurements of a single resistor, an LCR-circuit as well as an electrodynamic loudspeaker chassis [4] are shown in Fig. A.3. Each impedance function has been fitted with the corresponding theoretical expectation based on the particular component.

Applicability in the Lab-course Teaching Format

The presented idea combined with the low-cost and simple experimental setup appears suitable to serve as a lab course in the framework of *project-based learning* for undergraduate students in natural-sciences. The project demands and promotes a variety of skills, such as programming, hardware-control and data visualisation in PYTHON, electro-technical design and construction involving network theory (Kirchhoff's and Ohm's law), circuit soldering, knowledge about electrical components and a fundamental insight into data acquisition and digital signal processing and related experimental difficulties.
The project course may start with the question of 'how can we perform automated electrical measurements with non-dedicated scientific hardware, using our computer only'? The use of the input-output functionality of a computer sound card serves precisely this purpose. Based on this insight, an electronic network has to be designed

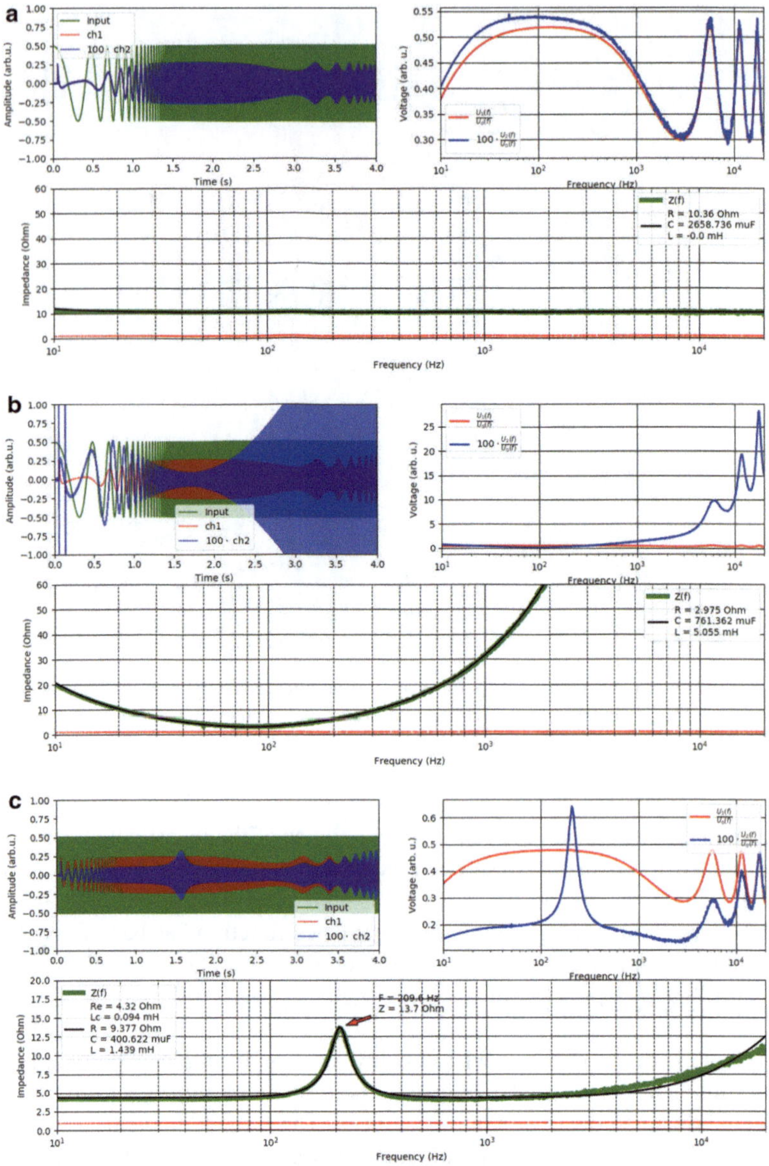

Fig. A.3 Provisional graphical user interface for impedance-spectroscopy measurements. Shown are three dataset windows corresponding to measurements of a single resistor (**a**), an LCR-series circuit (**b**) and a loudspeaker chassis (**c**), respectively. Each window contains three graphs. Up-left: Raw output (green) and input (red and blue) voltages over time. Up-right: Fourier-transformed voltage signals $U_{1,2}(f)$ as a function of frequency f. Down: Calculated impedance spectrum $Z(f)$ for the examined component. The fit functions are based on the theoretical electrical models of each of the three representative measurements

and built (for example the setup as described in the previous section) that allows for extracting the desired electrical value of the examined component. The next task may be to write a software code for accessing the input and output channels of the sound card. The completed setup and program code have to be tested and calibrated with known electronic components. For the subsequent measurements, a large variety of experiments can be performed by adopting the setup to the specific needs. A selection of measurements, all involving the field of materials science, is proposed:

- The characterisation of (self-built) electronic filter networks for their application in follow-up experiments.
- The characterisation of electrical transducers, receivers or other electronic components.
- The measurement of hysteresis curves on ferroelectric or ferromagnetic samples.
- A dielectric spectroscopy on different materials placed in between capacitor plates.
- A temporal analysis of corrosion processes by tracing the capacitance of a system evolving over time. Here, the formation of oxide layers increases the electrode's capacitance (this idea covers the topic of battery development).
- A mapping of the non-linear and/or non-reciprocal characteristics of electronic systems as for instance the $I - V$ curves of transistors or diodes.
- The excitation and detection of internal elastic resonances of materials in an *acoustic-resonance-spectroscopy* measurement.

The experimental course has been presented in our department, while the exact realisation and the sequence of experiments is still under debate with Dr. Martin B. Willeke (Head of lab courses at the DEPARTMENT OF MATERIALS, ETH ZURICH).

Appendix B
Design and Construction of Laboratory Hardware

I conceived and constructed different hardware for the daily use in our laboratories. The main constructions built for this work are presented in the following sections. I prefer the use of self-made experimental devices mainly because of three aspects:

- To avoid the use of 'black-boxes' in a setup and to facilitate an in-house service and repair.
- To benefit from the flexibility of implementing desired functionalities and to customise the equipment.
- To extend my knowledge and intuition of how to experimentally approach scientific tasks.

B.1 The Demagnetisation Setup

Demagnetising experiments, as explained in Sects. 3.3 and 5.1, are based on two main functionalities. First, a magnetic field has to be generated that surpasses the coercive fields or flipping fields of the material under consideration. This field has to be tunable in amplitude, such that a controlled decrease from maximal field to zero field is possible. Second, a fast reversal of the applied field has proved valuable [5–7] since it optimises the statistics and quasi-randomness of induced switching processes. Both considerations have been accounted for in the design of the self-built demagnetisation setup. Instead of a mere reversal of the applied field, a rotation of the field has been strived for to ensure an equal accessibility to differently aligned sublattices in the studied nanomagnetic arrays and to comply with any magnetic anisotropies. Several ideas have been considered as for instance the generation of a rotating magnetic field of controllable frequency and amplitude by an according modification of a stepper- or AC motor. The final realisation, however, is based on the idea to combine a sample rotation with an oscillating magnetic field. The sample can be rotated with a 0–24-VDC-driven commercially-available electro motor providing a voltage-controlled adjustable rotation frequency (here, set to 10^3 rpm). The magnetic

J. Lehmann, *Toroidal Order in Magnetic Metamaterials*, Springer Theses,
https://doi.org/10.1007/978-3-030-85495-9

field is generated by a stray-field transformer typically found in neon-lighting advertisement, see the right-hand side of Fig. B.5. It provides a strong magnetic field in a 2–3-cm-sized aperture, which is coupled to the power-grid frequency of 50 Hz. Maximum magnetic field strengths of 120 mT can be reached with the transformer used. A variable transformer is used for adjusting the input voltage for the stray-field transformer that defines the resulting magnetic field strength measured with a Hall sensor.
Combining a fast sample rotation with an in-plane oscillating magnetic field of adjustable amplitude allows for an efficient demagnetisation, see Fig. 5.1, of superior flexibility with respect to commonly employed setups that operate with a unidirectional field.

B.2 A Programmable Four-Axis Magnet

For a flexible magnetic-property measurement setup with an adjustable strength and orientation of the applied magnetic field, either a four-axis magnet or a vector magnet is required. My construction (designed and built together with Dr. Stefan Günther) strives mainly for the ability to generate vertically and horizontally aligned magnetic fields of several 100 mT flux density. Furthermore, its purpose of an operation in optical measurement setups requires a large aperture. Accordingly, a design was chosen that spatially separates the generation of the field (using large coils) from the position where they are applied (magnetic pole-shoes). The arrangement of four coils on a ring with flux guides in between to distribute the magnetic field towards the centred aperture is a reasonable choice. To allow for large magnetic field without 'memory' effects, the pole shoes of the magnet should be fabricated from a material of high saturation magnetisation, yet with a negligible remanence magnetisation. Soft (annealed) iron of high purity (here, 99.85 %) is an excellent choice with a saturation magnetisation above 2 T and a remanence magnetisation in the sub-mT to a few mT range [8, 9]. The wire for the coils requires a heat-resistant coating that allows for the high operational temperatures due to Joule' losses while operating. Since the coils will be wound by hand, a high current is preferred over a large number of windings to generate the desired field. Therefore, the wire has to allow for a continuous current load of— in the present case—above 10 A. Both considerations are pursued by using polyimide-coated copper wire (IRCE Poliflex 200) with a diameter of 1.8 mm. The conceived design has been simulated using FEMM, a software package based on finite elements approximating the achievable magnetic-field strengths and field homogeneities, see Fig. B.1. The iron core with an outer diameter of 180 mm has been sketched in a CAD program and fabricated with a CNC processing machine in the tool shop of our department.
To prevent oxidisation of the iron core, a heat-resistive polymer has been used for coating. The coils have been winded on freely rotating coil holders made of copper that have been placed around the outer rim of the iron core with the help of the materials-science student Bettina Tran. After winding, the coils have been locked in

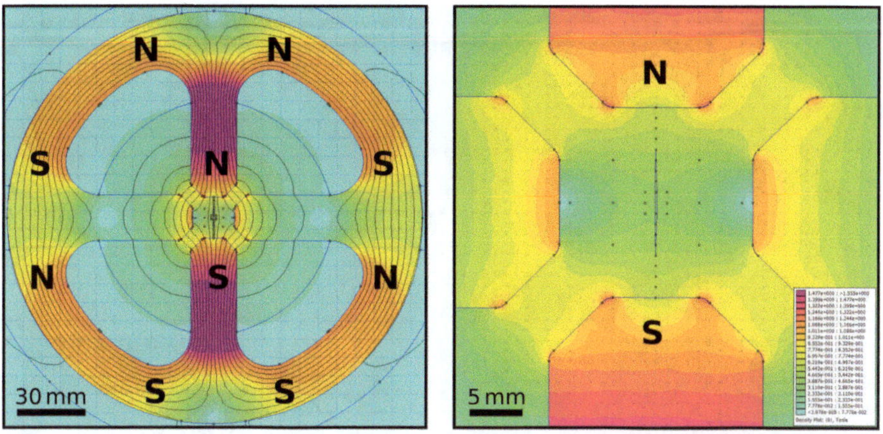

Fig. B.1 Finite elements simulation of the four-axis magnet. Sketched design of the iron core with the calculated magnetic-flux distribution for a vertically aligned magnetic field. The four field coils are placed on the sections of the outer rim. The poles are indicated with **N** (north) and **S** (south). The pole-shoe design requires a trade-off between a maximised magnetic-field amplitude on the one hand and a maximised field homogeneity on the other hand

place using a heat-resistive epoxy glue (LOCTITE EA 9492). The magnet is finally placed in a self-made height-changeable clamp that can be mounted on an optical table. Pictures of the described processes are shown in Fig. B.2.

For the pre-calibration of the magnet and for ensuring an identical field symmetry for all possible field orientations, the inductivity and resistance of each of the coils have been measured and adjusted separately. For connecting the four coils in a way to achieve different magnetic-field configurations, a computer-controlled unit has been developed using the NATIONAL INSTRUMENTS NI USB-6001, a programmable 14 bit input-output system. Eight output channels of this interface control simple transistor circuits that drive eight electrical relays. These relays can switch between sets of possible circuitries to choose between different modes of operation. This flexible circuit in combination with a computer-controlled bipolar power supply (KEPCO model BOP 20-20) can set a field geometry and strength on demand. A self-written LABVIEW program defines the field geometry and the durations and amplitudes of a field sweep, and simultaneously measures a field-dependent quantity such as a MOKE signal.

A test of the magnet, generating vertically and horizontally aligned magnetic fields, has been performed using a Hall sensor, see Fig. B.3. The measurement curves for both field directions match perfectly, confirming properly calibrated magnetic-field coils. While the maximum field strength in the linear regime is about $\pm 150\,\mathrm{mT}$, the remanence is as low as 5 mT. Note, however, that while operating in the linear regime only, the remanence drops below 1 mT. An overview of the parameters of the magnet is shown in Table B.1.

Fig. B.2 Construction of the four-axis magnet. Shown are process steps of the magnet fabrication starting with the raw CNC-machined iron core (upper left corner) to the final adjusted magnet in an optical setup (lower right corner)

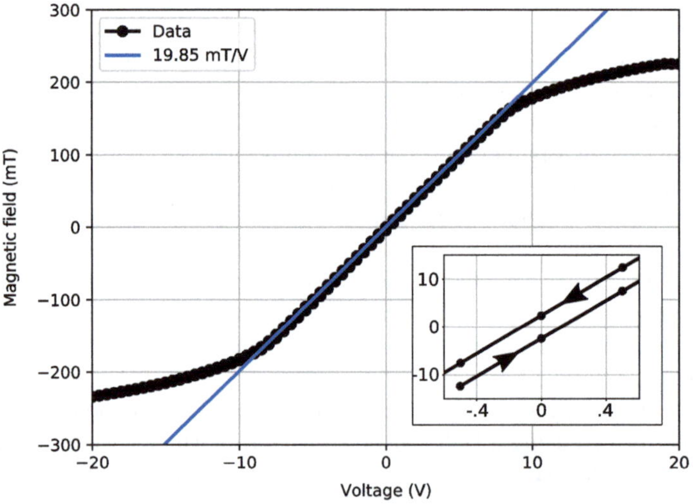

Fig. B.3 Magnetic-field curves in the horizontal direction. The magnetic field at the centre of the aperture has been measured with a Hall sensor. Voltages of up to ±10 V result in a linear response of the magnetic field until the saturation of the iron core flattens the curve. The inset highlights the magnet's maximum remanence of 5 mT after reaching saturation

Table B.1 Technical specifications of the four-axis magnet

Aperture	18 mm
Dimensions (h×w×d)	30–35×25×8 cm^3
Bore diameter	18 mm
Maximum current	20 A
Field strength	230 mT @ 20 A
Maximum remanence	5 mT
Typical remanence	< 1 mT
Serial resistance	1.6 Ω
Serial inductivity	5.4 mH
Field calibration	19.85 mT/V

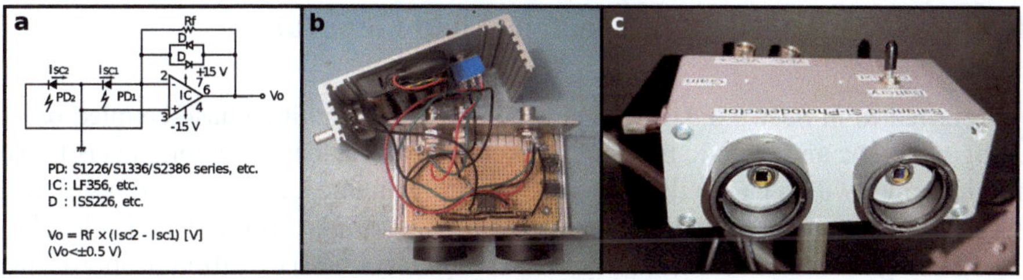

Fig. B.4 Design and construction of a balanced photo diode. a Sketch of the electrical circuit for balancing and amplifying the output of two photo diodes, adapted from Ref. [10]. **b** Open casing of the balanced photo diode. **c** Balanced photo detector in operation in the optical setup

B.3 Balanced Photo Diodes

For detecting small signals obtained with optical setups, two pathways can be taken in order to optimise the signal-to-noise ratio. One the one hand, the use of lock-in-detection schemes avoids an interference with parasitic signals that can result from, e.g., stray light. Lock-in detection therefore uses a periodically modulated excitation combined with a frequency-selective monitoring. On the other hand, a balanced detection scheme [10] uses a difference signal from two detectors and can overcome limitations resulting from an intensity detection with just one sensor. Such a balanced-detection system only involves a substitution of the detector with no further changes on the excitation side. It is therefore an effective and simple method to improve the sensitivity of a setup.

The technical principle of balanced detection is a reverse-parallel connection of two photodiodes such that their individual photo currents become subtracted (i.e. fully cancelled for an equal incident light intensity on both detectors). By doing so it is possible to avoid common-mode noise such as intensity fluctuations of the excitation light source by up to 50 dB [13]. The differential output of two diodes (here: HAMAMATSU S1226 silicon diodes with suitable sensitivity in the visible spectral range) typically serve as the input for a trans-impedance (current-voltage conversion) operational amplifier (OPAMP) (here, four different models have been proven suitable: LF- 356N, AD711, ADA4530- 1 and OPA227) with a current feedback via a resistor that, therefore, determines the gain, see Fig. B.4. The constructed balanced photo detector, see Fig. B.4, uses a potentiometer (Piher T-21, 0–10 MΩ) for adjusting the feedback current. Typical OPAMPs requires a stable and symmetric floating voltage input in the range of ±10 V. To comply with this requirement, a battery circuit (providing a stable symmetric output of ±3 V) and a self-built passively stabilised unregulated power supply providing ±10 V have been set up. A switch allows one to change between these two voltage sources to match the experimental requirements and the input-voltage specifications of the used OPAMP.

B.4 AC Magnetic- and Electric-Field Generators

As referred to in Sect. 6.2.3, different schemes for generating phase-coupled oscillating magnetic and electric fields have been conceived. The common idea is to use a transformer with a gap in the magnetic core to generate a concentrated oscillating magnetic field via an oscillating current in the transformer's primary coil. The electric field is generated simultaneously in the secondary coil. Two different approaches have been realised:

(1) Using the stray-field transformer introduced in Sect. B.1, 50-Hz-oscillating magnetic fields of up to 120 mT are generated within the yoke, see Fig. B.5a). The secondary coil of this high-voltage transformer provides a voltage of up to 6 kV. Between the 15-mm-separated electrodes of the sample holder (see Fig. B.5b) an electric field of $E \approx 4$ kV/cm is generated. The cross product of the two fields used for toroidal poling calculates to $\mu_0 H \times E = 480$ TV/cm, which is about twice the coercive field measured [11] for $LiCoPO_4$ of 250 TV/cm.
(2) Using the concept known as 'zero-voltage switching' (ZVS), oscillating electric and magnetic fields can be generated with much higher frequencies. The centrepiece is an LC resonance circuit consisting of a magnetic-field coil mounted on a gapped transformer core and a capacitor, both defining the operation frequency of the device. This circuit is driven by two transistors that are fed by a feedback signal recovered from a centre tap of the primary coil. The circuitry provides a pendulum-like chopping of the current passing through the transistors that is synchronised with the free oscillation frequency of the LC circuit. Related to the name of this circuit, it switches both transistors during the zero-voltage passage to avoid power dissipation and to allow for high work loads. The primary coil has been winded on a powder ferrite core specifically made for MHz-frequencies in which a centimetre-sized gap has been cut for accessing the induced magnetic field (see Fig. B.5c). The 1.2 μH inductivity of the coil in combination with a 220 μF high-voltage short-circuit-proof capacitor results in a resonance frequency of 300 kHz. Depending on the input voltage, here 51 VDC,

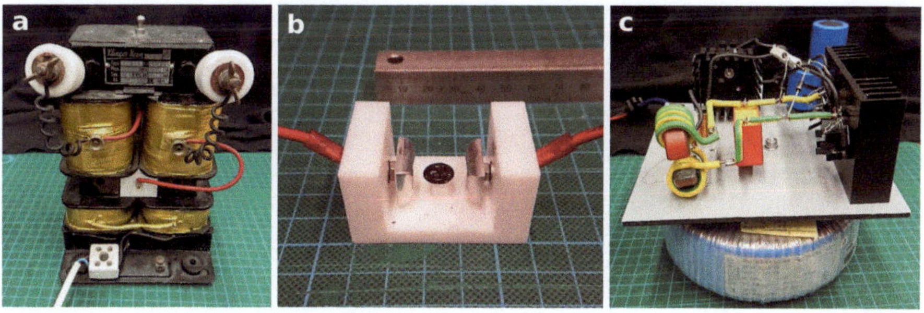

Fig. B.5 AC magnetic- and electric-field generators. **a** The stray-field transformer generating oscillating electric and magnetic fields of 50 Hz. The yoke of the transformer is replaced with the sample holder. **b** The sample holder made of poly-tetrafluorethylene (PTFE) and with aluminium electrodes. **c** The ZVS-based 300 kHz resonance circuit

the generated magnetic field strength reaches above 10 mT. A secondary coil allows for the simultaneous induction of a voltage of about 1 kV. Both fields can be applied simultaneously to the sample using the holder shown in the centred panel of Fig. B.5.

Appendix C
Additional Experiments on Nanomagnetic Arrays

The following sections contain additional measurements or calculations to provide a better insight and further informations for selected sections of the main text. The sections are not intended to be self-explaining, but require the corresponding background from the main text. Therefore, at the beginning of each section, a topic and page reference and is given to define the context.

C.1 In-situ Imaging of Magnetisation-Reversal Processes

The following content refers to Sect. 6.2.1 on page 115.
The scientific intention for the measurements that have been the starting point for the tip-assisted toroidal poling was to map—in-situ—microscopic changes in the magnetic configuration of a nanomagnetic array during magnetisation reversal. Therefore, the array has been imaged via MFM, using tips of low magnetic moment, see Table 3.1 on page 64. Simultaneously, a homogeneous in-plane magnetic field of variable strength has been applied.
A macrospin-based toroidal square array, see Fig. 4.1a, comprising of nanomagnets of size $200 \times 50 \times 20\,\text{nm}^3$, a lattice constant of 410 nm and 50 nm separation distances between neighbouring building blocks, has been the subject of the study. The array has been imaged with the fast scan direction along the [110] direction of the lattice and the slow scan direction parallel to $[1\bar{1}0]$. A magnetic field of 90 mT along [110] initiated the magnetic configuration of the array into a remanent-magnetised single-domain state, as imaged in the first few scanned lines as shown in the upper part of Fig. C.1b. To gradually destabilise the prepared magnetised state, the direction of the field has been reversed ($[\bar{1}\bar{1}0]$) and increased in steps of about 4 mT from 0–67 mT after approximatively every 1 μm (from top to bottom, see Fig. C.1). Therefore, the downwards-oriented vertical axis of Fig. C.1b corresponds to the spatially resolved magnetic configuration imaged during an increasing magnetic field, as listed.

J. Lehmann, *Toroidal Order in Magnetic Metamaterials*, Springer Theses,
https://doi.org/10.1007/978-3-030-85495-9

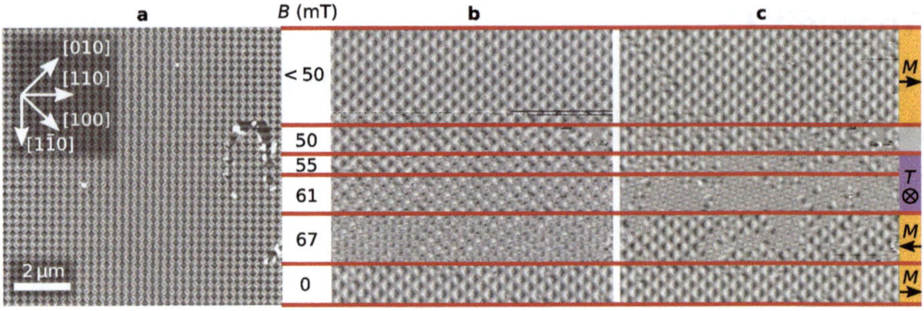

Fig. C.1 **Field-induced tip-assisted switching of magnetic configurations**. A section of the toroidal square array is scanned by a MFM tip during and after a magnetic field sweep. **a** Topography of the scanned area. **b** A magnetic single-domain state has been prepared by the application of a 90-mT-strong magnetic field along the [110]-direction of the array, which is above the coercive field. Subsequently, the array has been scanned line by line from top to bottom, with an increasing magnetic-field strength from 0–67 mT along [$\bar{1}\bar{1}$0], as indicated at the left side. Note that at field values between 55 and 67 mT, significant local reorientation events occurred as sudden changes in the observed magnetic configurations. **c** The same area is subsequently scanned without an applied magnetic field. Now, at the locations where reorientation events have happened, two striking observations can be made: First, a magnetic field of 55–61 mT results in a magnetic state with a net toroidisation oriented into the plane. Second, a magnetic field of 67 mT results in a local 180°-reversal of the initial in-plane magnetised state. Remarkably, the area scanned with no field applied (bottom lines) show no reorientation processes

Local modifications in the magnetic state of the array occurred after reaching a magnetic-field strength of about 55 mT. An increasing field strength lead to a rising number of switching events. The last few micrometre of the area have been scanned without an external magnetic field. A scan of this area demonstrates that the magnetic state in the un-scanned array has not been altered at all, see lower part of Fig. C.1b. Figure C.1c reveals the result of the magnetic-field-accompanied scan and displays the same area that has been scanned, here, without an external magnetic field. The magnetic state in areas that have been scanned during the application of a magnetic field of less than 50 mT is practically unchanged. A magnetic field of 50–55 mT applied during the scan induced a few local flips of the magnetisation. At a magnetic field of 61 mT applied during the scan, a net toroidisation of a particular sign ($-\vec{T}$) emerged. Furthermore, a field strength of 67 mT during the scan reversed the magnetisation direction of the array. Although a transition to the toroidal state and a complete reversal of the magnetic state have been achieved locally, no changes have been obtained in areas at which the external field superimposed with the tip field stood below the coercive field.

Figure C.2 provides more information on the key experiment of the tip-assisted toroidal poling procedure, see Sect. 6.2.1 on page 115. Together with the final state, it displays the changes of the magnetic configurations within the nanomagnetic array observed during the poling procedure.

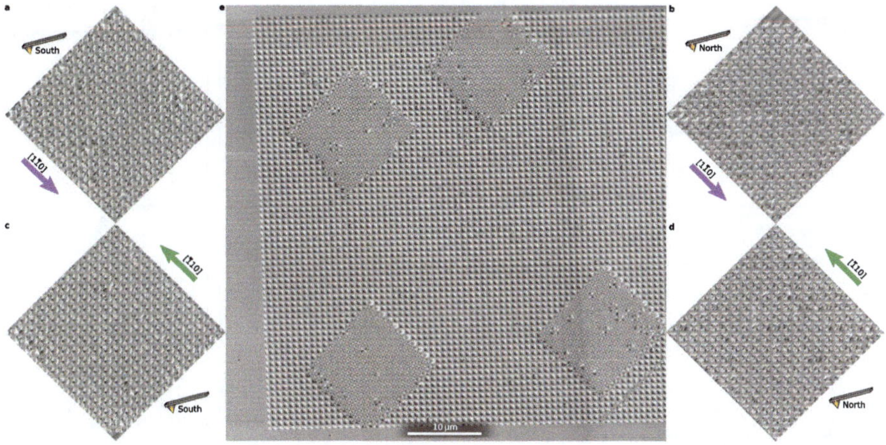

Fig. C.2 In situ imaging of tip-assisted toroidal poling. The centred MFM scan displays the entire toroidal square array after the application of poling fields of either sign at four selected areas. The regions indicated with **a–d** refer to the scans as shown in Fig. 6.3. Here, however, the scans reveal the changes in the magnetic configurations during the poling process, visible as sudden jumps in the magnetic state of individual macrospins. Again, different combinations of slow scan directions (violet and green block arrows) with out-of-plane tip-magnetisation directions (North and South) are shown

C.2 Micromagnetic Calculations to Re-Examine Toroidal Poling

The following content refers to Sect. 6.2.1 on page 115.

In order to test the micromagnetic calculations regarding their reliability and applicability, a set of parameter variations have been studied. The first variation considers the number of neighbouring magnets in the toroidal square array that have to be taken into account to get reliable results. In the calculations used for the respective chapter, seven neighbouring macrospins (the three neighbours within the unit cell as well as the four next-nearest neighbours of the adjacent unit cells) have been considered. The effect of a reduced number of neighbours is shown in Fig. C.3 and Table C.1. As an example, just the three nearest and next-nearest neighbours are considered and hence contribute to the calculated net magnetic field. No difference has been observed qualitatively for the net magnetic-field direction, while quantitatively, the calculated field values differ by less than 5 %.

In contrast, a change of the distances between macrospins and, thus, of the balance between $E_{\perp}$ and $E_{\parallel}$ is found to be a critical parameter. In order to remain in the desired switching mode, the relation between the two interaction energies has to fulfil $E_{\perp} > E_{\parallel}/2$, see Table C.1. However, a detailed analysis of this switching mode reveals,

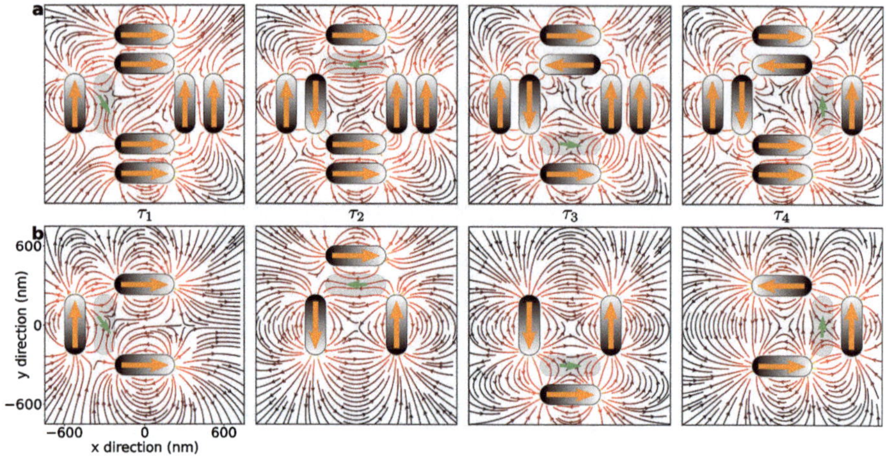

Fig. C.3 Details of local toroidal poling via crystal magnetic fields. Sequential tip-assisted 'activation' and reorientation of the four nanomagnets of a unit cell. The magnetic field exerted by the magnetic moments (orange block arrows) is indicated by red contours (see Fig. 3.12). The green block arrows represent the calculated direction of the net magnetic field averaged over the area of the respective nanomagnet above which the tip hovers at time τ_i. The direction of this net magnetic field determines whether the nanomagnet will reverse its magnetisation (τ_1, τ_2) or remain unswitched (τ_3, τ_4). **a** Consideration of seven neighbours for each time step. **b** Consideration of three neighbours for each time step. As a result, no qualitative difference between the two approximations is visible

that, although the scan across of the first unit cell yields just a full reversal of its magnetisation, toroidal switching into the opposite toroidisation state (as compared with the switching as described in the main text) is possible and will occur at all the subsequently scanned cells.

C.3 Vertex States in the Toroidal Square Array

This content refers to Sect. 5.3.1 on page 95 (Fig. C.4).

C.4 Hysteresis Measurements on Toroidal Square Arrays

This content refers to Sect. 5.3.1 on page 95 (Fig. C.5).

Table C.1 In-plane magnetic-field strength for different calculation parameters. Listed are the net magnetic-field values H_i at the position of a nanomagnet that is destabilised at time τ_i, see Fig. 6.1. The vector component along the easy axis of the respective magnet is emphasised and, accordingly, a minus sign indicates a magnetisation reversal. The magnetic-field values are normalised to largest calculated value within the selection. The first two lines compare a different approximation depths, yet with fixed lattice parameters. The second two lines compare different edge-to-edge distances between macrospins, parametrised with $d_\perp$ and $d_\parallel$. The third line indicates a complete reversal of the net magnetisation within the scanned unit cell. This result substantiates the stated requirement of $E_\perp > E_\parallel/2$ for the relation between interaction-energy terms in the system to achieve the described toroidal-poling scheme

Calculation parameter	H_1 (%)	H_2 (%)	H_3 (%)	H_4 (%)
7 neighbours, $d_\perp \mid d_\parallel$ = 70 nm | 70 nm	$\begin{pmatrix} +16.9 \\ \mathbf{-29.6} \end{pmatrix}$	$\begin{pmatrix} \mathbf{-75.4} \\ -2.25 \end{pmatrix}$	$\begin{pmatrix} \mathbf{+23.2} \\ -2.53 \end{pmatrix}$	$\begin{pmatrix} -1.15 \\ \mathbf{+22.8} \end{pmatrix}$
3 neighbours, $d_\perp \mid d_\parallel$ = 70 nm | 70 nm	$\begin{pmatrix} +18.8 \\ \mathbf{-30.7} \end{pmatrix}$	$\begin{pmatrix} \mathbf{-71.6} \\ -0.425 \end{pmatrix}$	$\begin{pmatrix} \mathbf{+21.7} \\ -0.675 \end{pmatrix}$	$\begin{pmatrix} +0.650 \\ \mathbf{+21.3} \end{pmatrix}$
7 neighbours, $d_\perp \mid d_\parallel$ = 100 nm | 30 nm	$\begin{pmatrix} +12.7 \\ \mathbf{-42.5} \end{pmatrix}$	$\begin{pmatrix} \mathbf{-74.7} \\ -1.05 \end{pmatrix}$	$\begin{pmatrix} \mathbf{-2.95} \\ -1.28 \end{pmatrix}$	$\begin{pmatrix} -14.3 \\ \mathbf{-36.0} \end{pmatrix}$
7 neighbours, $d_\perp \mid d_\parallel$ = 30 nm | 130 nm	$\begin{pmatrix} +31.7 \\ \mathbf{-21.7} \end{pmatrix}$	$\begin{pmatrix} \mathbf{-100} \\ -4.23 \end{pmatrix}$	$\begin{pmatrix} \mathbf{+62.8} \\ -4.15 \end{pmatrix}$	$\begin{pmatrix} -1.45 \\ \mathbf{+66.9} \end{pmatrix}$

C.5 Phase Diagram of the Toroidal Square Array

The following content refers to Sect. 5.3.3 on page 103.
A phase diagram has been calculated by my colleague Amadé Bortis using Monte-Carlo (MC) simulations based on the following Hamiltonian (Abstracted from the magnetic-dipole interaction) that describes the two dominant coupling parameters $J_\perp \equiv \frac{3\mu_0 m^2}{16\pi\sqrt{2}r_\perp^3}$ and $J_\parallel \equiv \frac{\mu_0 m^2}{4\pi r_\parallel^3}$ (see Fig. 3.11) in the toroidal square array.

$$\mathcal{H} = J_\perp \sum_{<i,j>_\perp} D_{ij} \cdot (\vec{m}_i \times \vec{m}_j) + J_\parallel \sum_{<i,j>_\parallel} \vec{m}_i \cdot \vec{m}_j. \tag{C.1}$$

Here, $\vec{m}_{i,j}$ parametrises the i, j-th perpendicular and parallel nearest neighbouring magnetic moments, respectively and $\vec{D}_{ij} = \vec{r}_i \times \vec{r}_j$ is a classical representation of the Dzyaloshinskii-Moriya vector with $\vec{r}_{i,j}$ as nanomagnet position vectors in the unit cell. The simulation calculates an annealing process and extracts the order parameter as well as the density of octupolar and quadrupolar subunits that form above the ordering temperature. The results are a main pillar of our recent publication, see Ref. [12] (Fig. C.6).

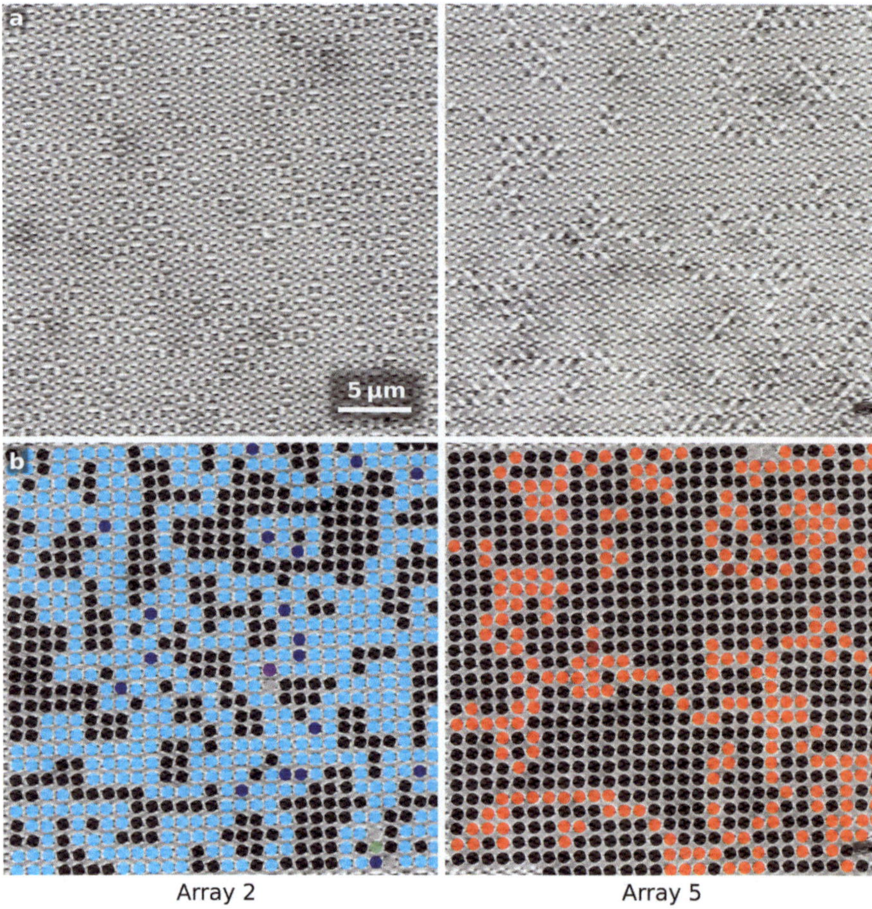

Fig. C.4 **Tuning microscopic interactions in toroidal square arrays**. Close-up of the vertex states from array 2 and array 5 shown in Fig. 5.7 on page 98. **a** MFM scans performed with a low-moment tip (see Table 3.1) showing approximately 30 × 30 unit cells. **b** Assignment of identified vertex states according to Fig. 5.6 and Table 5.1

C.6 Additional Magneto-Optical-Diffraction Measurements

The following content refers to Sect. 7.1 on page 135.
Further longitudinal DMOKE measurements (see Sect. 3.2.2 on page 69) have been performed on the $+k_y$-diffraction spot, as indicated in Fig. 7.2. The optical detection system uses a Glan-Thompson prism placed in front of a single photo diode to filter the initially set polarisation state and, thus, to provide polarisation sensitivity required for LMOKE. The four-axis magnet has been used to generate magnetic fields of $(-60, \ldots, 60)$ mT at the sample position applied along the $[\pm 100]$ direction

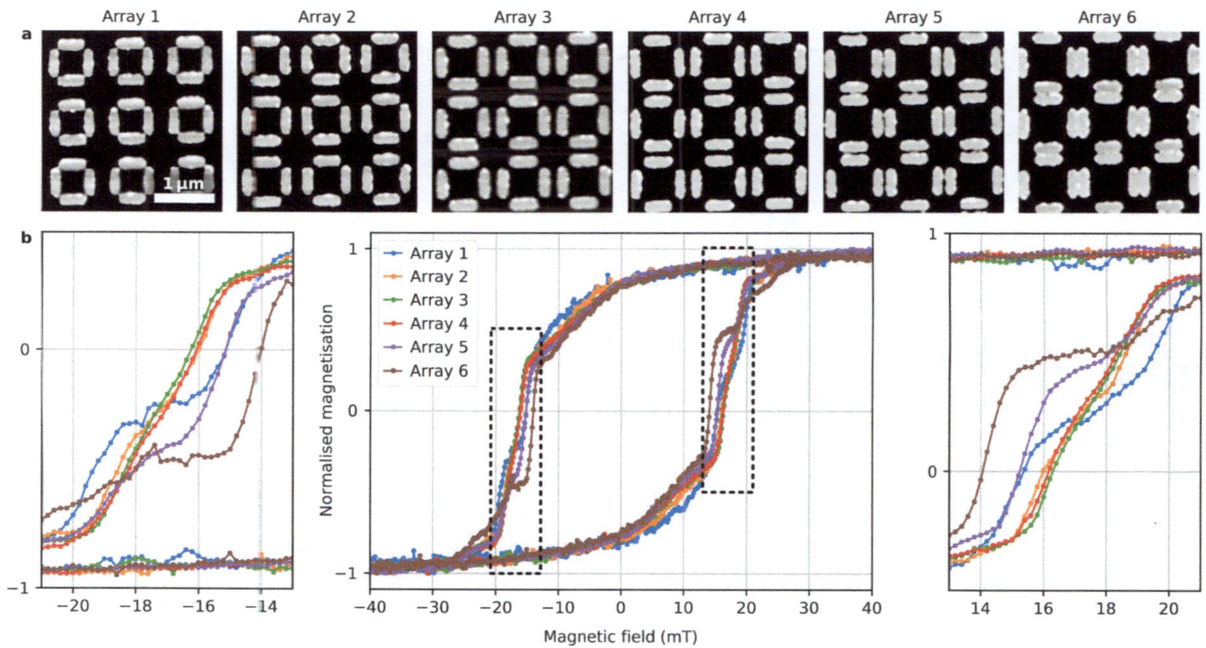

Fig. C.5 Variations in the coercive magnetic field as a function of microscopic interactions in toroidal square arrays. Shown are hysteresis curves measured on a series of toroidal square arrays using LMOKE (see Sect. 3.2.2 on page 65) using laser light of 405 nm wavelength, 10 mW power and 45° incidence with respect to the surface normal and a magnetic field along the structures diagonal. **a** AFM images of 3 × 3 unit cells of the different arrays. These arrays are identical to the ones evaluated in Sect. 5.3.1. **b** The complete hysteresis curves (centred panel) shows a very similar shape for each of the arrays, but with signatures of the magnetostatic coupling arising around the coercive magnetic field. The two panels on the left-hand side and the right-hand side, respectively, are magnified sections as indicated in the centred panel. The different switching characteristics are evident with coercive magnetic fields that differ by about 2 mT. Furthermore, while the Arrays 3 and 4 (corresponding to $E_\perp \approx E_\parallel$) show a relatively linear transition in the highlighted region, the Arrays 1 and 6 (corresponding to $E_\perp \gg E_\parallel$ and $E_\perp \ll E_\parallel$, respectively) show a clear two-step function indicating a metastable short-range ordered state emerging at the coercive magnetic field

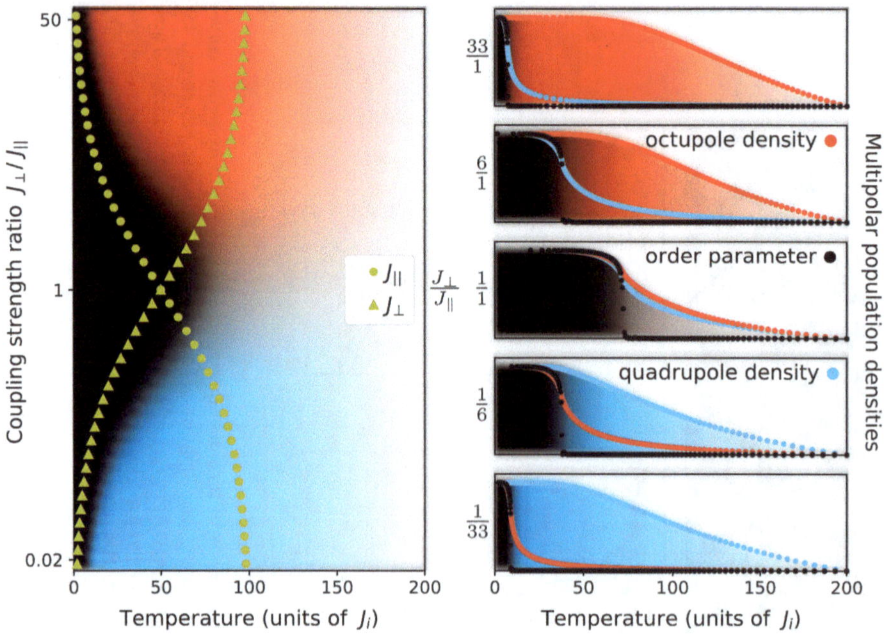

Fig. C.6 Phase diagram revealing short- and long-range order in toroidal square arrays. Left: MC-simulated phase diagram of the Hamiltonian, see Eqn. C.1, for a variety of coupling-strength ratios $J_\perp/J_\parallel$, each indicated with yellow triangles/circles. The formation of octupole and quadrupole subunits above the ordering temperature is evident (red and blue shading). The black shading indicates the long-range ordered ground state. Right: Formation of short- and long-range order upon simulated annealing for five exemplary choices of $J_\perp/J_\parallel$. The population densities are indicated by the intensity of the colour shading

of the array. This allows one to induce states of opposite net magnetisation as well as opposite net toroidisation within the hysteresis curve. An incident light polarisation along ±[101] and ±[010] (horizontal and vertical, respectively) resulted in zero signal amplitude, independent of the applied magnetic-field strength. This indicates an insensitivity of the measurement setting to the net magnetisation and net toroidisation at the chosen diffraction spot. However, for an incident polarisation of 45° ([111] direction) with respect to the sample normal a hysteresis curve has been measured, see Fig. C.7.

This hysteresis curve displays a similar shape as the specular magnetisation curve, as shown in Fig. 7.1, yet it differs regarding the relative signal strength of the plateaus. These plateaus, corresponding to the emergence, movement and annihilation of magnetic vortex states, both show a zero signal strength at the positive and negative coercive magnetic field. This support the implication that the polarisation state of the $+k_y$ reflex does not straightforwardly encode an information about the array's net magnetisation, which would lead to off-centred plateaus due to the field-induced bias of the vortex-core position.

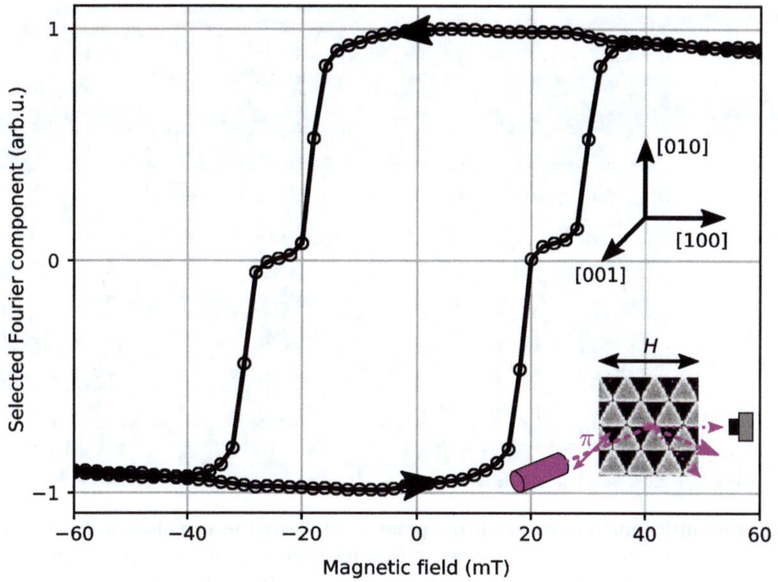

Fig. C.7 LDMOKE curve measured on a vortex-type toroidal array. Magnetic-field dependent polarisation rotation measured at the $+k_y$ optical-diffraction spot of an array of coupled magnetic vortices as shown in the inset and in Fig. 4.1d. A variable external magnetic field, generated by the four-axis magnet (see Appendix B.2), has been applied along [100]. The laser beam with 405 nm wavelength, 10 mW power and polarisation along $\pm$[111] was incident under 45° ([10$\bar{1}$]) with respect to the sample normal and focussed to a spot diameter of 30 μm at the sample to match the array size

C.7 Exposing Nanomagnetic Arrays to Intense Laser Pulses

The following content refers to Sect. 7 on page 133.

In collaboration with Dr. Oles Sendetskyi, a PhD candidate at Paul Scherrer Institute at that time, tests of the laser-induced damage threshold (LIDT) have been performed on artificial square-ice arrays (150 × 50 × 3 nm^3-sized permalloy nanomagnets on a silicon substrate), see Fig. 2.15 on page 38. A titanium-sapphire-laser-pumped optical parametric amplifier has been used that provides 100-fs-long pulses with 1 kHz repetition rate at a wavelength of 1200 nm (to avoid an above-bandgap excitation in the silicon substrate). Linearly polarised laser pulses of varying energy, incident under 45° with respect to the sample normal, have been focussed onto a spot diameter of about 50 μm (70 μm long axis) to match the typical size of individual nanomagnetic arrays. The illumination time has been set to 1 min.

In this configuration the structure remains entirely intact for laser-pulse energies of up to 250 nJ, corresponding to a fluence of $F = 9\,\mathrm{mJ/cm^2}$. The first hints of laser-induced damage have been observed to occur at a pulse energy of 700 nJ

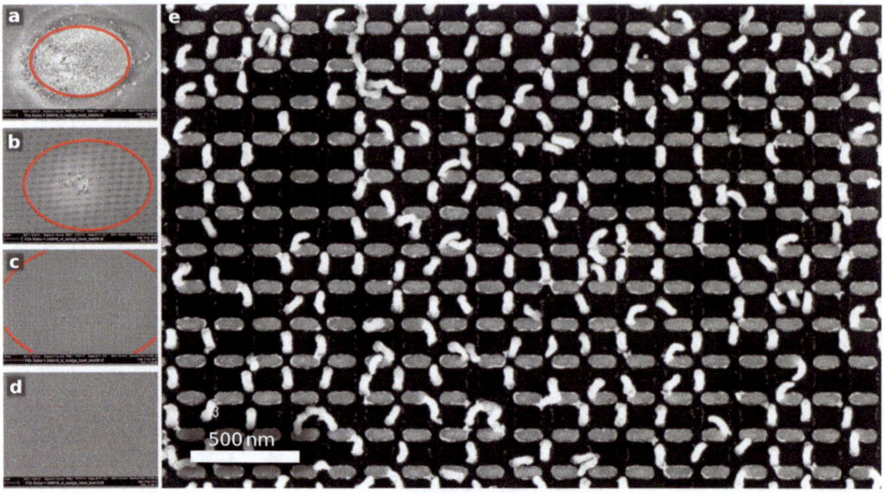

Fig. C.8 Sublattice melting by vertically polarised infrared laser pulses. **a–d** Laser-induced damage tests on an artificial square-ice-lattice (nanomagnet sizes of $150 \times 50 \times 3\,\text{nm}^3$). Each of the arrays has been exposed for 1 min to 100 fs-long vertically polarised laser pulses of 1200 nm wavelength and pulse energies of **a** 5 μJ, **b** 1.9 μJ, **c** 700 nJ and **d** 250 nJ. The red circle indicates the elliptical FWHM laser-beam profile of $50 \times 70\,\mu\text{m}^2$ (sub-figure **d** shows an area that is smaller than the laser-spot size). **e** magnified section from **c** exposed to a fluence of $25\,\text{mJ/cm}^2$. The observed melting of only vertically aligned nanomagnets indicates a site-selective heating due to a plasmonic-resonance effect

($F = 25\,\text{mJ/cm}^2$). At this pulse energy and by using a linear light polarisation a sublattice melting has been observed in the array. Here, only permalloy magnets with their long axis aligned parallel to the polarisation direction of the laser are warped and partly ablated. This a direct indication of a site-selective interaction of light with the nanomagnetic structure, possibly due to a plasmonic-resonance effect (Fig. C.8).

References

1. http://jp-minerals.org/vesta/en/. Online content, Accessed on 27 Jan 2020
2. Momma K, Izumi F (2011) VESTA 3 for three-dimensional visualization of crystal, volumetric and morphology data. J Appl Crystallogr 44:1272–1276. http://dx.doi.org/10.1107/S0021889811038970
3. Lehmann J et al (2019) Microdisplays as a versatile tool for the optical simulation of crystal diffraction in the classroom. J Appl Crystallogr 52(2):457–462. http://dx.doi.org/10.1107/S1600576719001948
4. Thorborg K, Unruh A, Struck CJ (2007) An improved electrical equivalent circuit model for dynamic moving coil transducers. Audio Eng Soc Conv 122:14
5. Ke X et al (2008) Energy minimization and ac demagnetization in a nanomagnet array. Phys Rev Lett 101(3):037205. http://dx.doi.org/10.1103/PhysRevLett.101.037205

6. Mengotti E et al (2008) Building blocks of an artificial kagome spin ice: photoemission electron microscopy of arrays of ferromagnetic islands. Phys Rev B 78(14):144402. http://dx.doi.org/10.1103/PhysRevB.78.144402
7. Nisoli C et al (2010) Effective temperature in an interacting vertex system: theory and experiment on artificial spin ice. Phys Rev Lett 105(4):047205. http://dx.doi.org/10.1103/PhysRevLett.105.047205
8. Jiles D (2015) Introduction to magnetism and magnetic materials, 3rd edn. CRC Press. http://dx.doi.org/10.1201/b18948
9. Danan H, Herr A, Meyer AJP (1968) New determinations of the saturation magnetization of nickel and iron. J Appl Phys 39(2):669–670. http://dx.doi.org/10.1063/1.2163571
10. https://www.hamamatsu.com/resources/pdf/ssd/si_pd_circuit_e.pdf. Online content, Accessed on 31 Jan 2020
11. Zimmermann AS, Meier D, Fiebig M (2014) Ferroic nature of magnetic toroidal order. Nat Commun 5(1):4796. http://dx.doi.org/10.1038/ncomms5796
12. Lehmann J et al (2020) Relation between microscopic interactions and macroscopic properties in ferroics. Nat Nanotech 15(11):896–900. http://dx.doi.org/10.1038/s41565-020-0763-9
13. https://www.rp-photonics.com/balanced_photodetection.html. Online content, Accessed on 31 Jan 2020

MIX
Papier aus verantwortungsvollen Quellen
Paper from responsible sources
FSC® C105338

If you have any concerns about our products,
you can contact us on
ProductSafety@springernature.com

In case Publisher is established outside the EU,
the EU authorized representative is:
Springer Nature Customer Service Center GmbH
Europaplatz 3, 69115 Heidelberg, Germany

Printed by Libri Plureos GmbH
in Hamburg, Germany